1
Topics in Organometallic Chemistry

Topics in Organometallic Chemistry

Forthcoming volumes:

Organometallic Bonding and Ractivity: Fundamental Studies
Volume Editor: J.M. Brown, P. Hofmann

Lanthanides: Chemistry and Use in Organic
Volume Editor: S. Kobayashi

Activation of Unreactive Bonds and Organic Synthesis
Volume Editor: S. Murai

Springer

Berlin
Heidelberg
New York
Barcelona
Budapest
Hong Kong
London
Milan
Paris
Singapore
Tokyo

Alkene Metathesis in Organic Synthesis

Volume Editor: A. Fürstner

With contribution by
S.E. Gibson, Y. He, A. Hoveyda, S.P. Keen,
L. Kiessling, N.P. King, M. Mori, K.C. Nicolaou,
J.H. Pawlow, R.R. Schrock, L.E. Strong,
D. Tindall, K.B. Wagener

 Springer

The series *Topics in Organometallic Chemistry* presents critical overviews of research results in organometallic chemistry, where new developments are having a significant influence on such diverse areas as organic synthesis, pharmaceutical research, biology, polymer research and materials science. Thus the scope of coverage includes a broad range of topics of pure and applied organometallic chemistry. Coverage is designed for a broad academic and industrial scientific readership starting at the graduate level, who want to be informed about new developments of progress and trends in this increasingly interdisciplinary field. Where appropriate, theoretical and mechanistic aspects are included in order to help the reader understand the underlying principles involved.
The individual volumes are thematic and the contributions are invited by the volumes editors.

In references Topics in Organometallic Chemistry is abbreviated
Top. Organomet. Chem. and is cited as journal

Springer WWW home page: http://www.springer.de

ISSN 1436-6002
ISBN 3-540-64254-4
Springer-Verlag Berlin Heidelberg New York

Library of Congress Cataloging–in–Publication Data
Alkene metathesis in organic synthesis / volume editor, A. Fürstner.
  p. cm. -- (Topics in organometallic chemistry ; 1)
  Includes bibliographical references.
  ISBN 3-540-64254-4 (alk. paper)
  1. Alkenes. 2. Ring-opening polymerization. 3. Metathesis (Chemistry)
4. Transition metal catalysts. I. Fürstner, Alois.
II. Series.
QD305.H7A45 1998
547′.412--dc21

© Springer-Verlag Berlin Heidelberg 1998
Printed in Germany

The use of general descriptive names, registered names, trademarks, etc. in this publication does not imply, even in the absence of a specific statement, that such names are exempt from the relevant protective laws and regulations and therefore free for general use.

Cover: Friedhelm Steinen-Broo, Pau/Spain; MEDIO, Berlin
Typesetting: Data conversion by MEDIO, Berlin

spin: 10551906   66/3020 - 5 4 3 2 1 0 – Printed on acid-free paper.

Volume Editor

Prof. Alois Fürstner
Max-Planck-Institut für Kohlenforschung
Kaiser-Wilhelm-Platz 1
D-45470 Mülheim an der Ruhr, Germany
e-mail: fuerstner@mpi-muelheim.mpg.de

QD305
H7A45
1998
CHEM

Editorial Board

Foreword

Organometallic chemistry is a well established research area at the interface of organic and inorganic chemistry. In recent years this field has undergone a renaissance as our understanding of organometallic structure, properties and mechanism has opened the way for the design of organometallic compounds and reactions tailored to the needs of such diverse areas as medicine, biology, materials and polymer sciences and organic synthesis. For example, in the development of new catalytic processes, organometallic chemistry is helping meet the challenge to society that the economic and environmental necessities of the future pose.

As this field becomes increasingly interdisciplinary, we recognize the need for critical overviews of new developments that are of broad significance. This is our goal in starting this new series Topics in Organometallic Chemistry.

The scope of coverage includes a broad range of topics of pure and applied organometallic chemistry, where new breakthroughs are being achieved that are of significance to a larger scientific audience. Topics in Organometallic Chemistry differs from existing review series in that each volume is thematic, giving an overview of an area that has reached a stage of maturity such that coverage in a single review article is no longer possible. Furthermore, the treatment addresses a broad audience of researchers, who are not specialists in the field, starting at the graduate student level. Discussion of possible future research directions in the areas covered by the individual volumes is welcome. Finally, the coverage is conceptual, focussed and concise, presenting the most significant results and the underlying principles that are emerging. Thus where appropriate, the authors are encouraged to include theoretical and mechanistic aspects. Indeed, one of the first volumes, Organometallic Bonding and Reactivity: Fundamental Studies, edited by John Brown and Peter Hofmann, will be a cornerstone of the series.

It seems appropriate that the first volume of this series, Alkene Metathesis in Organic Synthesis, edited by Alois Fürstner, covers one of the most exciting and dynamic areas of modern organometallic research. Several volumes on other areas, where significant breakthroughs are being achieved, will be published soon. Furthermore we are convinced that organometallic chemistry is such a fertile and expanding research field that there will be many new topics in the future that will deserve coverage.

John M. Brown, Oxford
Pierre Dixneuf, Rennes
Alois Fürstner, Mülheim
Louis S. Hegedus, Fort Collins
Peter Hofmann, Heidelberg
Paul Knochel, Marburg
Tobin J. Marks, Evanston
Shinji Murai, Osaka
Manfred Reetz, Mülheim
Gerard van Koten, Utrecht   September 1998

Preface

Olefin metathesis has its roots in polymer chemistry, but it remained a laboratory curiosity in organic synthesis for decades, simply because most of the early metathesis catalysts are more or less incompatible with polar functional groups. The advent of metal alkylidene chemistry completely changed this situation: various complexes of this type were found to be well defined, single component (pre)catalysts that are distinguished by a high performance as well as by a remarkable tolerance towards polar groups. The impact of this discovery can hardly be overestimated: it led to the design of user-friendly tools for all types of olefin metathesis reactions which, in turn, enable highly imaginative syntheses of complex target molecules and opened up new vistas for advanced polymer chemistry and materials science.

This monograph of the new Springer series *"Topics in Organometallic Chemistry"* is not intended to provide a comprehensive treatise of metathesis. Even more so, I guess that such an endeavour would be in vain at the present time due to the explosive growth in the number of publications, which substantiates the rapidly increasing importance and popularity of this transformation. However, the volume collects a series of review articles highlighting some of the most exciting advancements that have been reached in recent years. They may enchant the reader, will serve as a valuable source of information, and hopefully will stimulate further creative work in this timely field of research. I am indebted to all fellow chemists who generously contributed their expertise and knowledge to this project.

Mülheim, September 1998   Alois Fürstner

Contents

Olefin Metathesis by Well-Defined Complexes of Molybdenum and Tungsten

Richard R. Schrock

Olefin metathesis is a catalytic process whose key step consists of a reaction between an olefin and a transition metal alkylidene complex. Some of the best understood alkylidene catalysts contain Mo or W in its highest possible oxidation state, along with "supporting" alkoxide ligands and an imido ligand. Recent advances in our understanding of these catalysts from a fundamental perspective, and in ring opening metathesis polymerization (ROMP) and ring closing metathesis (RCM), in particular, are reviewed.

Keywords: Metathesis, Molybdenum, Tungsten Alkylidene, Ring-opening, Ring-closing.

1
Introduction

Olefin metathesis is a catalytic process whose key step is a reaction between an olefin and a transition metal alkylidene complex, usually M=CHR (Eq. 1) or M=CH$_2$, in a 2+2 fashion to give an unstable intermediate metalacyclobutane ring [1]. All possible reactions of this general type are reversible, possibly nonproductive, and in competition with one another, so the overall result depends heavily on relative rates, and in the case of formation of volatile or insoluble products, displacement of equilibria as those products form.

$$M=CHR_1 + R_1CH=CHR_2 \longrightarrow \begin{matrix} M-CHR_1 \\ | \quad | \\ R_2HC-CHR_1 \end{matrix} \longrightarrow M=CHR_2 + R_1CH=CHR_1$$

$$\text{cis or trans} \quad (1)$$

Much of the proof for the mechanism of this reaction has been provided by studies that involve isolated alkylidene or metalacyclobutane complexes, first of Ti [2–4] and Ta [5], and then of Mo and W [6, 7]. In all such species the metal is in its highest possible oxidation state if the alkylidene ligand is viewed as a dianion. Alkylidene complexes are also called carbene complexes, especially for later transition metals in which the metal–carbon double bond is not as polarized $(M(\delta+)=C(\delta-))$ as it is in an early transition metal complex, and the metal is not in its highest possible oxidation state. Some later metal complexes are active metathesis catalysts, especially ruthenium complexes of the type $Ru(CHR)Cl_2$ $(PR'_3)_2$ [8–11]. Later transition metal carbene complexes have a higher tolerance than "d^0" alkylidene complexes toward oxygen and protic reagents, even water, consistent with their less polar nature and the less "oxophilic" nature of the later transition metal. The trade-off is that the reactivity of ruthenium complexes toward more hindered or substituted C=C bonds is lower than the reactivity of Mo or W alkylidene species. In "classical" catalyst systems, side reactions that lead to isomerization of the C=C bond are sometimes observed, usually long after metathesis equilibrium has been established [1]. Isomerization of the C=C bond has now been observed in metathesis reactions catalyzed by the "well-defined" or "well-characterized" Mo and W complexes reviewed here [12, 13], although it is safe to say that double bond isomerization is so rare that it is not a serious concern.

Several fundamental types of metathesis reactions for monoolefins or diolefins are shown in Eqs. 2–5.

Coupling
$$2\ RCH=CH_2 \longrightarrow RCH=CHR + CH_2=CH_2 \quad (2)$$

ADMET
$$y\ CH_2=CH(CH_2)_xCH=CH_2 \longrightarrow \overline{[CH(CH_2)_xCH]}_y + y\ CH_2=CH_2 \quad (3)$$

$$\text{RCM} \quad (4)$$

$$+ CH_2=CH_2$$

$$\text{ROMP} \quad (5)$$

Removal of a product (e.g., ethylene, Eq. 2) from the system could dramatically alter the course and/or rate of a desired metathesis reaction, since ethylene reacts with an alkylidene complex to form a methylene $(M=CH_2)$ complex, which is the most reactive (and also the least stable) of the alkylidene complexes. Of potentially greater interest than homocoupling (Eq. 2 [14]) is cross-coupling between two terminal olefins, especially if one cannot be homometathesized. Coupling reactions involving dienes — only α,ω dienes will be used as examples — can lead to linear and cyclic dimers, oligomers, and, ultimately, linear or cyclic polymers (ADMET [15–29]; Eq. 3). In general ADMET is favored in highly concentrated solutions or in neat substrate, while cyclization is favored at low concentrations. If the diene couples intramolecularly to give a cyclic alkene, the process is called ring-closing metathesis (RCM; e.g., Eq. 4) [10]); RCM has become an important area of research in the last few years, in particular in the context of its application to organic chemistry. Cyclic olefins can be opened and oligomerized or polymerized (ring opening metathesis polymerization or ROMP [2, 30]; Eq. 5). When the alkylidene catalyst reacts more rapidly with the cyclic olefin (e.g., a norbornene or a cyclobutene) than with a C=C bond in the growing polymer chain, then "living ROMP" can result, i.e., there is little termination during or after a polymerization reaction. A wide variety of block copolymers therefore can be prepared [30]. If one also considers that *cis* or *trans* olefins can be formed in any metathesis step, that all possible alkylidenes can form, that alkylidenes and olefins can vary widely in their reactivity, and that all reactions are reversible to a greater or lesser extent, then the potential complexity of some even ostensibly simple metathesis systems becomes apparent. The reactions shown in Eqs. 2, 4, and 5 will be the focus of this article, although articles covering these and other reactions from other perspectives can be found elsewhere in this volume.

Several circumstances can dramatically reduce the turnover frequency of a metathesis reaction. For example, the least reactive alkylidene (e.g., a disubstituted alkylidene) may be the exclusive or dominant species because of its generally low reactivity, but it will not react readily with the most reactive olefin at a significant rate. A second example of a limited turnover frequency is when a metalacyclobutane intermediate is formed that is relatively stable toward loss of an olefin. A third example is simple competitive inhibition of olefin binding by some donor base, either in an intramolecular fashion in an alkylidene complex, or intermolecularly. Finally, any reaction that destroys the alkylidene (e.g., reaction with the carbonyl function, oxygen, water, protic solvents, etc.) will obviously limit longevity and therefore turnover number. Therefore important issues in systems that involve Mo or W alkoxide complexes in particular are the purity of solvents, substrates, and atmosphere. Systems in which the rate of destruction of catalytic intermediates is low compared to a productive reaction are obviously the most desirable, and some trade-off in terms of purification of solvents and substrates and catalyst loading is usually necessary. However, a negative result for a single reaction under a given set of conditions could have many causes, and in general it is wise to check a known reaction under the same conditions at the same time.

An alkylidene complex of Mo or W also can react with an acetylene to generate an intermediate metalacyclobutene intermediate and a new alkylidene via rearrangement of the metalacyclobutene (Eq. 6). The new alkylidene may be further involved in metathesis reactions, even though its disubstituted nature generally reduces its reactivity compared to the starting M=CHR complex. (Intramolecular versions of reactions involving such species will be dramatically faster than intermolecular versions, and are known to be viable.) Repetition of reactions analogous to that shown in Eq. 6 can lead to oligomerization or polymerization of alkynes to give polyenes [31–34]. Reactions involving acetylenes also can be controlled by "fine tuning" the reactivity of an isolable alkylidene initiator [35–38]. A reaction that involves an acetylene also can be employed in a "tandem" fashion, i.e., as an intermediate step in some olefin metathesis reactions [39, 40]. This reaction and more extensive "cascade" or "relay" reactions could prove useful in certain circumstances.

$$
\text{M=CHR + R'C}\equiv\text{CR'} \longrightarrow \text{M}\underset{R'}{\overset{R'}{<}}\!\!\!=\!\text{CHR} \tag{6}
$$

The vast majority of the "well-defined" Mo and W catalysts that will be discussed here in detail contain alkoxide ligands. The sequence of events that led to the realization that alkoxide ligands are compatible with metathetical reactions has been recorded elsewhere [6, 7]. Alkoxide ligands must be relatively bulky in order to prevent dimerization or intermolecular decomposition reactions. The electronic variability of an alkoxide ligand then allows the reactivity of the metal to be "tuned" over a wide range. The effect of changing the alkoxide from $OCMe_3$ to $OCMe(CF_3)_2$ is enormous, especially since two alkoxides are bound to the metal in a metathesis catalyst in general. Catalysts that contain other supporting ligands have been prepared [1], although their efficiency generally has not been documented extensively. (A later section of this article will be devoted to some of these species.) Other ligands that are present throughout the metathesis process (e.g., imido or oxo) can also play important roles. However, the use of terms such as "spectator" or "ancillary" to describe ligands that are bound to the metal throughout the metathesis process is inappropriate, as the only truly ancillary ligand is one that is not bound to the metal during the metathesis process. A labile base that must be lost to provide a coordination site on the metal falls into this category.

In this chapter I will cover only "well-defined" or "well-characterized" compounds. Results will be included that have appeared since reviews in 1991 on alkylidene and metalacyclobutane complexes [41] and in 1993 on ring-opening metathesis polymerization [30], but an overview of prior results that are especially relevant to olefin metathesis in particular will also be included. (An excellent and comprehensive text also has been published recently [1].) The terms "well-defined" or "well-characterized" originally were meant to imply that the alkylidene complex is isolable and is essentially identical to that in a catalytic reaction except for the identity of the alkylidene. These terms have been watered down from time to time in the literature, even to the point where they are used to describe a catalyst that is formed from a "well-characterized" transition metal "precursor complex," but whose identity actually is not known. In this article I will adhere to the original definition. The alkylidene complexes at least should be observable in solution, and often some variation should have been isolated, structurally characterized, and shown to initiate metathesis without addition of any "activator." The most that will be required will be ligand dissociation. Some of the species discussed herein have been shown to be stable in dilute solution for weeks or even months [12], even though they cannot be isolated in the solid state, which is good evidence that bimolecular decomposition reactions can be slowed at high dilution to the point of being nearly irrelevant. If the reactivity of the catalyst toward olefins is high enough, then metathesis reactions can be sustained for a significant period of time. Contributions by other researchers to the chemistry of "d^0" alkylidene complexes can be found in previous review articles [2, 5, 30, 41–44]. Key contributions will be mentioned here again from time to time. Several "well-defined" Mo and Ru alkylidene complexes, as well as the precursor to a family of Mo catalysts, are now available commercially from Strem Chemicals.

The value of a 100% active, "well-defined" species with known structure and activity is now becoming appreciated. The possibility of stoichiometric or even catalytic reactions resulting from other metal species in solution was a significant disadvantage of "classical" metathesis catalysts where only a minuscule amount of a necessarily very active species was responsible for catalysis [1]. However, the development of simple and expedient "recipes" that comprise an in situ synthesis of a "well-defined" catalyst, or a species close to it, may be desirable in some circumstances [45]. The advantages of an in situ synthesis are speed and simplicity and the fact that the yield of desired species need not be as high as in a typical synthesis where the product must be isolated, *as long as* no reagents or reaction products other than the desired alkylidene interfere with the desired metathesis reaction. However, a specific metathesis reaction by a given type of alkylidene complex may be difficult to ensure if it is not the most reactive species present. The risk of forming small amounts of a highly reactive, but unselective catalyst is real in all systems, but probably highest in circumstances where the true catalyst has not been isolated and purified. Documented cases of improved catalyst performance upon purification (usually recrystallization) of a well-defined catalyst have now been reported in the literature [46].

2
Methods of Synthesizing Mo and W Alkylidene Complexes

Mo and W alkylidene complexes usually are prepared from M(VI) dialkyl complexes by some variant of the α hydrogen abstraction reaction (Eq. 7; other ligands omitted) [5, 41].

$$M(CH_2R)_2 \longrightarrow M=CHR + CH_3R \tag{7}$$

The source of the alkyls is usually some main group alkylating agent, and the groups that are replaced on M to form a dialkyl complex usually are halides. When the alkyl group contains one or more β protons, then the predominant reaction usually is a related β hydrogen abstraction reaction to give a metal olefin complex in which the metal formally has been reduced by two electrons. Recently it has been demonstrated in tantalum chemistry that α abstraction can be faster than β abstraction in sterically crowded circumstances that limit the β proton coming close enough to the metal to be activated for β abstraction [47–49]. However, in general it is believed that β abstraction reactions predominate over α abstraction reactions. β Abstraction is not possible when neopentyl, neophyl (CH_2CMe_2Ph), trimethylsilylmethyl, or benzyl ligands are employed. The rate of α abstraction appears to be highest with neopentyl or neophyl ligands and the resulting alkylidene is the most stable of the primary alkylidenes. The rate of α abstraction in dimethyl complexes is many orders of magnitude slower, and the resulting methylene species are the most susceptible to bimolecular decomposition reactions by many orders of magnitude. For all of these reasons neopentylidene or neophylidene ligands are often found in well-defined metathesis initiators. α-Hydrogen abstraction can be induced by light, although there are few examples of photochemically induced α-abstraction to give a stable alkylidene complex in high yield [50]. Nevertheless, catalysts generated in situ via photoinduced α abstraction in some cases have been found to be useful initiators in metathesis reactions [51].

An alternative to α abstraction is to add the "pre-formed" alkylidene from another element to a M(IV) species, or to employ a "masked" alkylidene. The first example of a reaction of the first type to give a high oxidation state terminal alkylidene was the reaction between $Ta(\eta^5\text{-}C_5H_5)_2(PR_3)Me$ and alkylidene phosphoranes to give $Ta(\eta^5\text{-}C_5H_5)_2(CHR)Me$ (R=H or Me) species [53]. Tungsten complexes of the type $W(CHCMe_3)(O)(PR_3)_2Cl_2$ also were first prepared by transfer of an alkylidene from tantalum to tungsten [54]. A recent example is the reaction between an alkylidenephosphorane and $W(NAr)(PR_3)_3Cl_2$ complexes to give complexes of the type $W(CHR)(NAr)(PR_3)_2Cl_2$ ($Ar=2,6\text{-}R_2C_6H_3$; Eq. 8) [52].

$$W(NAr)(PR_3)_3Cl_2 + R'_3P{=}CHR \xrightarrow[-PR'_3]{-PR_3} W(NAr)(CHR)(PR_3)_2Cl_2 \qquad (8)$$

$$\tag{9}$$

In the latter category is a variation of a reaction that was first reported by Binger [55] for preparing Ti vinylalkylidene complexes, an example of which is shown in Eq. 9. This type of reaction has been used to prepare tungsten vinyl alkylidene complexes by treating $W(NAr)(PR_3)_3Cl_2$ complexes (R=alkyl or OMe) with 3,3-disubstituted cyclopropenes [56]. Similar reactions between $W(O)(PR_3)_3Cl_2$ (R= alkyl or OMe) and 3,3-diphenylcyclopropene or ketalcyclopropene yield analogous oxo alkylidene complexes [57, 58]. The oxo and imido dihalide complexes are analogous to bisphosphine tungsten alkylidene complexes prepared earlier

[5], and which were found not to be long-lived metathesis catalysts. However, in some cases the halides can be replaced by bulky alkoxides and long-lived active catalysts thereby prepared.

3
Oxo Alkylidene Complexes

The first isolated alkylidene complexes of Mo or W that showed significant metathesis activity were pseudo-octahedral oxo alkylidene complexes of tungsten, $W(O)(CH-t-Bu)(PR_3)_2Cl_2$ [54]. It was noted at that time that "$W(O)(CH-t-Bu)(O-t-Bu)_2$" could be prepared, although it proved too unstable to characterize. In general oxo alkylidene complexes that are capable of reacting with an olefin readily are too unstable toward bimolecular decomposition reactions to be isolated, or even observed in solution at a concentration sufficient for NMR detection. Well-characterized and metathetically active oxo alkylidene complexes in fact are still rare. Oxo alkylidene complexes are probably the type of alkylidene species present (in dilute solution) in many classical metathesis systems, and also in systems in which alkoxide or phenoxide ligands, along with an oxo ligand, are present in the starting metal W(VI) complex [45]. Oxo complexes also tend to be relatively reactive (e.g., toward alkylating agents) in undesirable ways, a fact that leads to low yields of oxo alkylidene species. Nevertheless, oxo alkylidene complexes could prove to be important in future systems, if ways can be found to prepare them in a straightforward manner and in a respectable yield, and to control their reactivity.

Rare examples of stable and metathetically active oxo alkylidene complexes are syn-W(CH-t-Bu)(O)(OAr)$_2$(L) complexes (Ar=2,6-Ph$_2$C$_6$H$_3$; L=PMe$_3$ or PPh$_2$Me), prepared by treating W(CH-t-Bu)(O)(PMe$_3$)$_2$Cl$_2$ [54] with two equivalents of KO-2,6-Ph$_2$C$_6$H$_3$. (The syn rotamer is that in which the alkylidene substituent points toward the oxo ligand.) An X-ray study of syn-W(CH-t-Bu)(O)(OAr)$_2$(PPh$_2$Me) showed it to be a distorted trigonal bipyramid in which the oxo ligand, the neopentylidene ligand, and the oxygen of one phenoxide ligand all lie in the equatorial plane, a structure that is analogous to that found for adducts of imido alkylidene complexes [41]. It is proposed that the phosphine occupies the apical position where an olefin coordinates to give an "olefin/alkylidene" intermediate that has a trigonal bipyramidal geometry; therefore the phosphine must be labile enough to generate a four-coordinate oxo alkylidene complex rapidly. Studies involving competitive binding of a base in the position where an incoming olefin is presumed to bind can be found in the literature [37, 59].

4
Imido Alkylidene Complexes

The primary reason for attempting to synthesize imido alkylidene complexes, e.g., W(NR)(CH-t-Bu)(O-t-Bu)$_2$, was the belief that the appropriate imido ligand will block bimolecular decomposition reactions more effectively than an oxo ligand by not functioning as a bridging ligand. The relatively bulky (and inexpensive) 2,6-diisopropylphenylimido ligand was chosen precisely for those reasons, and because the 2,6-diisopropylphenoxide ligand had been employed successfully in acetylene metathesis systems [60].

The synthesis of the first imido alkylidene complex to which alkoxides could be attached is shown in Eq. 10 (Ar=2,6-i-Pr$_2$C$_6$H$_3$) [61]. It was fortuitous that a proton could be transferred from the arylamido ligand to the alkylidyne [60] ligand, since it is now known that the *reverse* is a potential complicating feature of certain imido ligands in certain circumstances, especially more basic *alkyl*imido ligands (see later). A later synthesis of a bistriflate analog of the dichloride complex [62] began with W(NAr)$_2$Cl$_2$(1,2-dimethoxyethane) (Eq. 11) and employed two equivalents of a neopentyl reagent to produce the alkylidene via α hydrogen abstraction in a hypothetical bistriflate dineopentyl intermediate. The less expensive neophyl (CH$_2$CMe$_2$Ph) ligand is usually employed instead of the neopentyl ligand today.

$$(10)$$

$$(11)$$

Neutral, four-coordinate complexes of the type W(CH-t-Bu)(NAr)(OR)$_2$ can be prepared readily from W(CH-t-Bu)(NAr)Cl$_2$(dme) or W(CH-t-Bu)(NAr)(triflate)$_2$(dme) as long as the alkoxide is relatively bulky (OR=O-t-Bu, OCMe$_2$(CF$_3$), OCMe(CF$_3$)$_2$, OC(CF$_3$)$_2$(CF$_2$CF$_2$CF$_3$), or O-2,6-R$_2$C$_6$H$_3$) [62–64]. The activity of such species for the metathesis of ordinary internal olefins (e.g., *cis*-2-pentene) appeared to peak for the OCMe(CF$_3$)$_2$ species. In some cases intermediate tungstacyclobutane complexes were stable enough to be observed or even isolated. Unsubstituted tungstacycles formed in the presence of excess ethylene proved to be especially stable. The tungstacyclobutane complex, W(CH$_2$CH$_2$CH$_2$)(NAr)[OCMe(CF$_3$)$_2$]$_2$, was shown to be a trigonal bipyramid with the imido group in an axial position and the WC$_3$ ring in the equatorial plane. On the basis of this work it was proposed that the rate of reaction of alkylidene complexes with olefins correlated directly with the electron-withdrawing ability of the alkoxide. However, it was clear that unsubstituted tungstacyclobutane complexes were much more stable toward loss of an olefin than substituted tungstacyclobutane complexes, presumably largely for steric reasons, and that the rate of metathesis of terminal olefins therefore could be attenuated as a consequence of formation of a relatively stable metalacycle. The situation became more complex when it was discovered that either trigonal bipyramidal (TBP) *or* square pyramidal (SP) tungstacyclobutane complexes could form, and that in some cases (OR=OCMe$_2$(CF$_3$) or OAr) a *mixture* of interconverting SP and TBP species (Eq. 12) could be observed.

$$(12)$$

In one case the lowest energy form depended upon the nature of the metalacycle, i.e., W[CH$_2$CH(R)CH$_2$](NAr)(OAr)$_2$ is a square pyramid when R=t-Bu, but a trigonal bipyramid when R=SiMe$_3$. It was proposed on the basis of kinetic studies that square pyramidal metalacycles are relatively stable toward loss of an olefin because the WC$_3$ ring is further from an "olefin/alkylidene" transition state than is the WC$_3$ ring in a trigonal bipyramidal metalacycle. For that reason complexes that contain relatively electron-withdrawing alkoxides (which are usually trigonal bipyramids) will lose an olefin more readily than those that contain relatively electron-donating alkoxides (which are usually square pyramids), in spite of the fact that the metal is "more electrophilic" when electron-withdrawing alkoxides are present. The fascination with tungstacycles began to wane when it was realized that metalacycles of molybdenum are dramatically less stable, even unsubstituted molybdacyclobutane complexes, and therefore in many cases molybdenum may be more active as a consequence of little or no catalyst being "trapped" in the metalacyclobutane form. This is the reason why Mo(CHR)(NAr)(OR')$_2$ catalysts generally are favored in metathesis reactions where ethylene is generated.

One of the most important early findings, although the significance was not realized at the time, was that *both syn and anti rotamers* of W(CHSiMe$_3$)(NAr)(OAr)$_2$ could be observed (Eq. 13) and that they interconvert on the NMR time scale ($\Delta G^{\ddagger} \approx 12$ kcal mol^{-1}). (The alkylidene is required to lie in the N-W-C plane in order to form a W=C π bond.)

$$(13)$$

In virtually all other W(CHR)(NAr)(OR')$_2$ complexes only the *syn* alkylidene rotamer is observed readily [63]. It was not clear at the time why rotamers could be observed in this particular case and why they interconverted readily. Later it was shown that the reactivities of certain *syn* and *anti* species could differ by many orders of magnitude and that the rates of their interconversion also could differ by many orders of magnitude as OR was changed from O-t-Bu to OC-Me(CF$_3$)$_2$. Therefore in *any* system of this general type two different alkylidene rotamers could be accessible (although both may not be observable), either by rotation about the M=C bond, or as a consequence of the metathesis reaction itself. The presence of *syn* and *anti* rotamers further complicates the metathesis reaction at a molecular level, and at least in ROMP reactions (see below) in important ways. The apparent ease of interconversion of *syn* and *anti* rotamers in phenoxide complexes could be an important feature of systems in which access to both *syn* and *anti* rotamers must be assured (see later).

A facile route to Mo(CH-t-Bu)(NAr)(OR)$_2$ species [65] was developed that began with [NH$_4$]$_2$[Mo$_2$O$_7$] and yielded Mo(CHCMe$_2$Ph)(NAr)(triflate)$_2$(dme) via Mo(NAr)$_2$Cl$_2$(dme) in a procedure analogous to that shown in Eq. 11. Therefore a wide variety of molybdenum complexes of the type Mo(CHCMe$_2$Ph)(NAr)(OR)$_2$ became available for study. (The triflate complex, a precursor to virtually any bisalkoxide species, is available commercially from Strem Chemicals.) Molybdenum complexes became the focus of most research since it was believed that they might be more tolerant of functionalities such as the carbonyl group than tungsten complexes, and that molybdacyclobutane complexes would be less stable than tungstacyclobutane complexes and therefore less often "sinks" that would sequester a catalyst in an inactive form.

Adducts of M(CH-t-Bu)(NAr)(OR)$_2$ complexes were prepared and studied as models for the initial olefin adduct [66] in an olefin metathesis reaction [67]. PMe$_3$ was found to attack the CNO face of *syn*-M(CH-t-Bu)(NAr)(OR)$_2$ rotamers to give TBP species in which the phosphine is bound in an axial position (Eq. 14).

anti CNO adduct   syn CNO adduct

(14)

It should be noted that this adduct is chiral and that the opposite enantiomer would result upon addition of the phosphine to the other CNO face, which corresponds basically to addition of the phosphine to the other side of the M=C bond. Predictably, the base is bound most strongly when the alkoxide is electron-withdrawing (e.g., OCMe(CF$_3$)$_2$), and is more strongly bound to W than to Mo. A relatively stable CNO adduct of the *syn* rotamer forms first when OR' is OCMe(CF$_3$)$_2$, as it is virtually the only rotamer present ($K_{eq} \approx 10^3$), and as was shown later (see below), the rate of conversion of the *syn* to the *anti* rotamer when OR=OCMe(CF$_3$)$_2$ is relatively slow (k$_{s/a}$≈10^{-5} s^{-1}). However, the CNO adduct of the *anti* rotamer is the thermodynamic product. It is believed to form via loss of PMe$_3$ from the *syn* adduct, followed by slow rotation of the alkylidene to give unobservable *anti*-M(CH-*t*-Bu)(NAr)(OR)$_2$, and trapping of the relatively reactive *anti* form by PMe$_3$. The CNO adduct of the *syn* rotamer is believed to be formed less readily because of the developing steric interaction between the R substituent on the *syn* alkylidene and the isopropyl groups on the aryl ring of the NAr ligand, which lies in the trigonal plane in the *syn* TBP olefin adduct. In the *anti* adduct only the alkylidene proton interferes with one isopropyl group in the Ar ring in the five-coordinate adduct. Although these studies do not prove that an olefin adds to the CNO face, as opposed to the COO or NOO faces, several theoretical studies that also explore a variety of other issues also suggest that is the case [68–74]. The stability of the base-free *syn* form over the base-free *anti* form has been attributed in part to an "agostic" M(CH$_\alpha$) interaction [5, 75] in the *syn* rotamer, one that is not possible in the *anti* rotamer.

A detailed investigation of alkylidene rotation rates in Mo(CHCMe$_2$Ph)(NAr) (OR)$_2$ complexes (where OR=OCMe$_2$(CF$_3$), OCMe(CF$_3$)$_2$, OC(CF$_3$)$_3$, or OC(CF$_3$)$_2$ (CF$_2$CF$_2$CF$_3$)) in thf or toluene produced some startling results [76, 77]. Values for k$_{a/s}$ were found to vary by up to seven orders of magnitude (at 298 K), the smallest values for k$_{a/s}$ being found in complexes that contain the most electron-withdrawing alkoxides in thf as a solvent, while equilibrium constants (K$_{eq}$= k$_{a/s}$/k$_{s/a}$) at 25°C were found to vary by two orders of magnitude at the most. Values for k$_{s/a}$ at 298 K could be calculated from k$_{a/s}$ and K$_{eq}$ and found to vary by up to six orders of magnitude in the same general direction as k$_{a/s}$. The main conclusion was that the rate of interconversion of *syn* and *anti* rotamers was "fast" for *t*-butoxide complexes (k$_{s/a}$≈1 s^{-1} at 298 K) and "slow" for hexafluoro-*t*-butoxide complexes (k$_{s/a}$≈10^{-5} s^{-1} at 298 K). To a first (and qualitative) approximation, when the metal is relatively electron-rich the alkylidene that has rotated by 90° can be stabilized by the orbital that lies in the N-M-C plane (Eq. 15).

anti   syn   (15)

When the metal is relatively electron-poor, that orbital is energetically more closely matched with the energy of a p orbital on the imido nitrogen atom and therefore is involved primarily in forming the pseudo triple bond to the imido ligand. The ease of rotation also varies to a significant degree with the nature of the imido and alkylidene ligands. For example, although there is little difference between the rates of alkylidene ligand rotation in hexafluoro-*t*-butoxide complexes that contain N-2,6-i-Pr$_2$C$_6$H$_3$ or N-2,6-Me$_2$C$_6$H$_3$ ligands, the alkylidene ligand in an analogous N-2-*t*-BuC$_6$H$_4$ complex rotates ~1500 times faster. The postulate is that the *unsymmetrically substituted* phenylimido ligand is bent to some extent in the ground state, thereby making the transition state containing a "bent imido" ligand shown in Eq. 15 more readily accessible. Similar phenomena that lead to bending of alkoxide or phenoxide ligands could also significantly alter the accessibility of various transition states, although at present it is not clear how such phenomena could be detected and documented.

5
Variations and Other Types of Alkylidene Complexes

A potentially useful variant of the synthesis of some Mo(CHR)(NAr)(OR')$_2$ complexes (R=CMe$_3$ or CMe$_2$Ph) consists of addition of two equivalents of a relatively acidic alcohol (R'OH=a fluorinated alcohol or phenol) to Mo(NAr)(N-t-Bu)(CH$_2$R)$_2$ [78, 79], a variation of the reaction of that type that was first reported in 1989 [80]. The more basic t-butylimido ligand is protonated selectively. This synthesis avoids the addition of triflic acid to Mo(NAr)$_2$(CH$_2$R)$_2$ to give Mo(CHR)(N-t-Bu)(triflate)$_2$(dimethoxyethane), the universal precursor to any Mo(CHR)(NAr)(OR')$_2$ complex. Unfortunately, the method does not appear to succeed when R'OH does not have a relatively high pK$_a$.

One might have thought that the imido ligand could be varied to a significant degree in complexes of the type Mo(CHR)(NR")(OR')$_2$, and that catalytic activity also could be controlled in that manner. While some variations in R" have been successful (2-t-BuC$_6$H$_4$, 2-i-PrC$_6$H$_4$, 2-PhC$_6$H$_4$, 2-CF$_3$C$_6$H$_4$, 1-admantyl) [81–83], sterically bulky aryl-substituted imido ligands (2,6-Me$_2$C$_6$H$_3$ or 2,6-i-Pr$_2$C$_6$H$_3$) have been the most successful and relatively innocent so far. Surprisingly, complexes of the type Mo(CHR)(N-t-Bu)(OR')$_2$ in general are not stable and well-behaved, perhaps in part because of the relatively basic nature of the t-butylimido ligand, making the alkylidyne tautomer, Mo(NH-t-Bu)(CR)(OR')$_2$, relatively close in energy and kinetically accessible (see immediately below). An effort to prepare Mo=NCMe(CF$_3$)$_2$ alkylidene species by the usual synthetic route noted earlier failed [84].

Ostensibly minor variations of a synthetic procedure sometimes can have significant consequences. For example, substituting KOCMe(CF$_3$)$_2$ for LiOCMe(CF$_3$)$_2$ is believed [85] to lead to formation of the alkylidyne complex shown in Eq. 16 instead of the known [83] Mo(CH-*t*-Bu)(NAd)[OCMe(CF$_3$)$_2$]$_2$ (Ad=adamantyl). A proton is likely to be transferred before formation of the final product, since it has been known for some time that both W(CH-*t*-Bu)(NAr)[OCMe(CF$_3$)$_2$]$_2$ and W(C-*t*-Bu)(NHAr)[OCMe(CF$_3$)$_2$]$_2$ are stable species that cannot be interconverted in the presence of triethylamine [41]. In such circumstances the nucleophilicity of the alkoxide ion and the nucleophilicity and acidity of the alcohol formed upon deprotonation of the alkylidene will be crucial determinants of whether the imido nitrogen atom is protonated at some stage during the reaction. At this stage few details are known about side reactions in which amido alkylidyne complexes are formed.

$$\text{Mo(CH-t-Bu)(NAd)(dme)(triflate)}_2 \xrightarrow[\text{- 2 KOTf - dme}]{2 \text{ KOCMe(CF}_3)_2} \qquad (16)$$

Complexes of the type Mo(CHCMe₂Ph)(NAr)(diolate) where the diolate is [R₄tart]²⁻, [BINO(SiMe₂Ph)₂]²⁻, or [Bipheno(t-Bu)₄]²⁻ (see Scheme 1) have been prepared by adding the diolate to Mo(CHCMe₂Ph)(NAryl)(triflate)₂(dme) [86]. [R₄tart]²⁻ was enantiomerically pure (+), while [BINO(SiMe₂Ph)₂]²⁻ was used in its racemic form. All were found to control the structure of a ROMP polymer in a dramatic manner, as discussed in more detail in a later section. Complexes that contain the [Bipheno(t-Bu)₄]²⁻ ligand are technically not permanently chiral, as rotation about the aryl-aryl C-C bond in the ligand could lead to interconversion of enantiomers. Analogous complexes that contain the [BINO(aryl)₂]²⁻ (aryl= phenyl, 2-methylphenyl, 2,6-dimethylphenyl, or 3,5-diphenylphenyl) or [Bipheno(t-Bu)₄Me₂]²⁻ ligands (see below) were also prepared and shown to yield high *cis*, isotactic polymers via ROMP of substituted norbornenes [87]. The [Bipheno(t-Bu)₄Me₂]²⁻ ligand cannot racemize readily and therefore yields complexes that are essentially permanently chiral at the metal center. H₂[BINO(SiMe₂Ph)₂] also was prepared in enantiomerically pure (-) form and used to prepare enantiomerically pure complexes. Related W complexes of the type W(X)(biphenoxide)Cl₂(THF) (X=O or N-2,6-Me₂C₆H₃) have been prepared by adding the free biphenol to W(X)Cl₄ [88] and have been employed as precursors to what are presumed to be catalysts of the type W(CHR)(X)(biphenoxide) formed upon alkylation with Et₂AlCl.

Molybdenum imido alkylidene catalysts that contain the enantiomerically pure diolate derived from a *trans* disubstituted cyclopentane (see Scheme 1) also have been prepared and employed in ring-closing reactions (see later) [89, 90]. This diolate forms a nine-membered MoO₂C₆ ring, in contrast to the seven-membered MoO₂C₄ rings in binaphtholate or biphenolate catalysts. No diolate complex is known where a five-membered MoO₂C₂ ring is formed, presumably because of the relatively exposed and strained nature of the diolate oxygen atoms in such a circumstance. However, such species could be stabilized by a base such as trimethylphosphine. (See below for an example containing the [1,2-(Me₃SiN)₂C₆H₄]²⁻ ligand.)

A bulkyl silsesquioxane has been used to prepare metathetically active Mo complexes [91]. A silsesquioxane mimics the manner in which a metal might be attached to silica in a heterogeneous metathesis system. The conjugate acid of the silsesquioxane appears to have a relatively high pK$_a$, a hint that the silsesquioxane itself may be relatively electron withdrawing. Silsesquioxane complexes were found to be highly active catalysts in metathesis reactions.

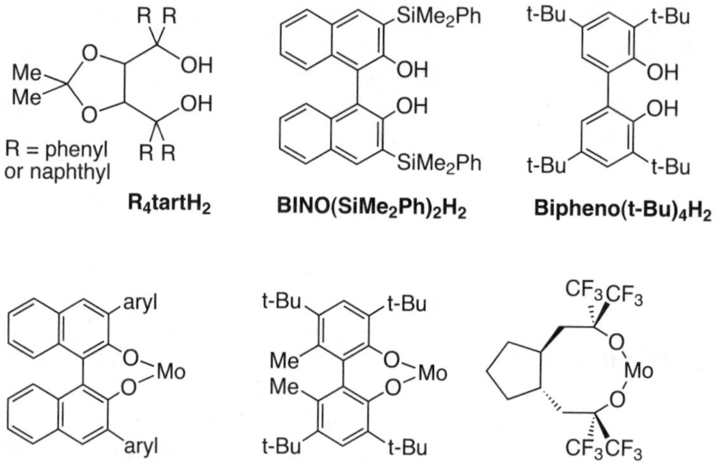

Scheme 1. Ligands found in catalysts of the type Mo(CHR)(NR')(diolate)

A trigonal bipyramidal trimethylphosphine adduct of a tungsten imido alkylidene complex that contains a substituted orthophenylenediamine ligand, $W(CH-t-Bu)(NPh)[1,2-(Me_3SiN)_2C_6H_4](PMe_3)$, has been reported [92]. Polymerization of norbornene could be observed, along with formation of *syn* and *anti* resonances ascribable to the propagating alkylidene complexes. Trimethylphosphine is believed to behave as a competitive inhibitor, analogous to what has been observed for a tungsten bisalkoxide imido alkylidene complex used to polymerize cyclobutene [59].

Molybdenum imido alkylidene complexes have been prepared that contain bulky carboxylate ligands such as triphenylacetate [35]. Such species are isolable, perhaps in part because the carboxylate is bound to the metal in an η^2 fashion and the steric bulk prevents a carboxylate from bridging between metals. If carboxylates are counted as chelating three electron donors, and the linear imido ligand forms a pseudo triple bond to the metal, then bis(η^2-carboxylate) species are formally 18 electron complexes. They are poor catalysts for the metathesis of ordinary olefins, because the metal is electronically saturated unless one of the carboxylates "slips" to an η^1 coordination mode. However, they do react with terminal acetylenes of the propargylic type (see below).

A variety of tungsten and molybdenum oxo and imido alkylidene complexes that contain a trispyrazolylborate (Tp) ligand have been prepared [93–98]. Species such as $W(O)(CHPh)(Tp)Cl$ are relatively air-stable, moisture-stable, and thermally stable. When combined with 1 equivalent of $AlCl_3$ these precursors yield active ADMET and ROMP catalysts. However, no data were offered in support of the claim that the active catalysts in these metathesis reactions also are air-stable and moisture-stable. Complexes of the type $TpW(CHCMe_2R)(CH_2CMe_2R)(NAr)$ are protonated by the noncoordinating Bronsted acid $[H(ether)_2]$ $\{B[3,5-(CF_3)_2C_6H_3]_4\}$ to give the cationic alkylidene complexes $[TpW(NPh)$

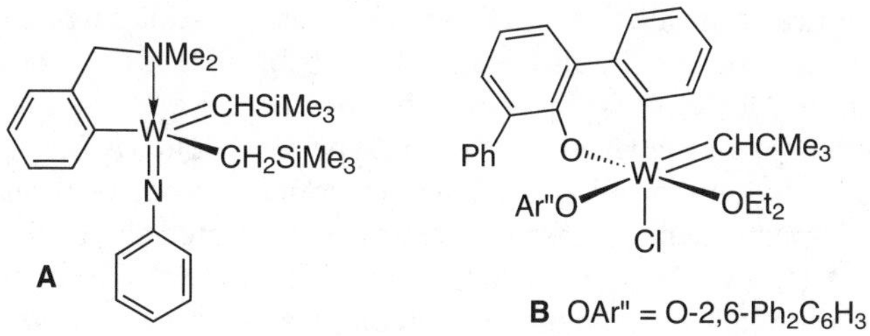

Scheme 2. Two tungsten alkylidene catalysts that contain a metal-aryl bond

(CHCMe$_3$)(Solvent)]{BAr[3,5-(CF$_3$)$_2$C$_6$H$_3$]$_4$}. Cationic alkylidene complexes in general are rare, although they are likely to be the most active species in the systems pioneered by Osborn and coworkers [99, 100].

A variety of catalytically active five-coordinate tungsten oxo and imido alkylidene complexes also have been prepared that contain some donor amine or pyridine linked either to the imido ligand or to a phenyl ligand bound to the metal (A, Scheme 2) [101–105]. Such species show metathesis activity (e.g., ROMP of norbornene), but there does not appear to be any proof that the integrity of the initiator is maintained.

The majority of Mo and W catalysts contain a sterically bulky aryl-substituted imido ligand. However, Basset and his group have prepared tungsten alkylidene complexes that contain a metalated phenoxide ligand, but no oxo or imido ligand (B, Scheme 2) [106–108]. Such species are related to catalysts prepared and studied by Osborn and coworkers [99, 100]. They have been used in a variety of metathesis reactions, although the true nature of the catalyst present during metathesis has not been confirmed. Heppert [109] has prepared related species such as (R$_2$BINO)(t-BuO)$_2$W(CHR) by treating (t-BuO)$_3$W≡CR complexes with 1,1'-bi-2-naphthols (H$_2$R$_2$BINO, R=Me,Br,Ph). An analogous reaction between (t-BuO)$_3$W≡CPh and 2 equiv of H$_2$Me$_2$BINO gave (Me$_2$BINO)$_2$W(CHPh). ROMP of norbornene was observed with complexes of this general type, but activity was only poor to modest unless the complex was activated with (e.g.) AlCl$_3$. Under these conditions at least one alkoxide is likely to be exchanged for a chloride, and highly active cationic species may be formed.

Molybdenum dinitrosyl complexes with the general formula Mo(NO)$_2$(CHR)(OR')$_2$(AlCl$_2$)$_2$ have been found to be active in a variety of metathesis reactions [110]. New alkylidenes could be identified. Variations such as Mo(NO)$_2$(CHMe)(RCO$_2$)$_2$ also are known [111]. Complexes of this type are believed to be more reduced than typical "d^0" species discussed here, although they appear to be much more active as metathesis catalysts than typical "Fischer-type" carbene complexes.

In the last several years tungsten alkylidyne complexes [60], W(CCMe$_3$)(CH$_2$CMe$_3$)$_3$ and W(CCMe$_3$)Cl$_2$(dimethoxyethane) in particular, have been a source of alkylidene complexes bound to oxide surfaces as a consequence of protonation of the alkylidyne ligand by a surface-bound hydroxyl group [112–114].

The exact nature of the alkylidenes formed on various oxide surfaces is still uncertain, as is the nature of the alkylidenes responsible for the often observed metathesis activity. $Mo(N)(CH_2CMe_3)_3$ also has been employed as a precursor to a surface-bound species believed to be of the type $Mo(NH)(CHCMe_3)(CH_2CMe_3)$ (O_{surf}) [115]. Although the alkylidene carbon atom could not be observed in solid state NMR spectra, which is typical of surface supported alkylidenes, reaction with acetone to give 2,4,4-trimethylpent-2-ene quantitatively confirmed the presence of the reactive neopentylidene complex. Such species would initiate various metathesis reactions when prepared on partially dehydroxylated silica.

6
Cross-Coupling Reactions

One might suppose to a first approximation that the cross-coupled metathesis product and the homocoupled products of two different terminal olefins would be formed in statistical yields, and that olefins that cannot be homocoupled could not be cross-coupled. However, Crowe showed that cross-coupled products could be formed in far greater than statistical yields [116]. Styrene and substituted styrenes could be cross-metathesized with a variety of functionalized terminal olefins to give unsymmetrical olefin products with >95% *trans* selectivity in up to 90% yield using the $Mo(CHCMe_2Ph)(N-2,6-i-Pr_2C_6H_3)[OC-Me(CF_3)_2]_2$ (MoF6) catalyst (1%). It was then found that acrylonitrile could be involved in a relatively selective cross-coupling reaction, even though the homocoupled dinitrile could not be formed [117]. Finally, allyltrimethylsilane was shown to yield a mixture of *cis* and *trans* cross-coupled products in yields that varied between ~40% and ~85% [118]. Bromides, nitriles, esters, and TBS-protected ether derivatives were all employed successfully to give cross-coupled products. It is clear from these studies that failure of an olefin to be homometathesized is *not* a good indication that it cannot be involved in any metathesis reaction under any circumstances.

The reaction shown in Eq. 17 [119] emphasizes the efficiency of certain cross-couplings, even in sterically crowded circumstances in the presence of relatively reactive functionalities. When Mo and Ru catalysts were compared, it was found that Mo generally required less time and gave higher yields, but is sensitive to hydroxy and acetate functionalities. Ru failed in cross-couplings involving acrylonitrile. It is likely that the Mo catalyst loadings could be reduced significantly in view of the high yields reported. In some cases 75% cross-coupled product could be obtained, even though each α olefin separately would form the homocoupled product in high yield (Eq. 18).

$$\text{BnO} \underset{3}{\diagdown} \diagup \diagdown + 1.3 \diagup \diagdown \diagdown_{\text{t-Bu}} \xrightarrow[\text{2 h, 70\%}]{5\% \text{ MoF6}} \text{BnO} \underset{3}{\diagdown} \diagup \diagdown \diagdown \diagdown_{\text{t-Bu}} \qquad (18)$$

The tandem use of asymmetric allylboration to give enantiomerically pure homoallylic alcohols followed by cross-metathesis of homoallylic silyl ethers with p-substituted styrenes has been reported [120] (Eq. 19). Exclusively *trans* cross-coupled products were formed in 50–75% yields.

$$\overset{\text{OSiMe}_2(\text{t-Bu})}{R \diagup \diagdown \diagup \diagdown} + Ar \diagup \diagdown \diagup \xrightarrow[\text{CH}_2\text{Cl}_2,\ 3\ \text{h}]{1\% \text{ MoF6}} R \diagup \overset{\text{OSiMe}_2(\text{t-Bu})}{\diagdown \diagup \diagdown} Ar \qquad (19)$$

Allyltriphenylstannane has been employed successfully in cross-couplings with a variety of terminal olefins containing (inter alia) ketal, ester, and cyano groups [121]. Only the MoF6 catalyst appeared to work at a reasonable rate in most cases. Both E and Z products were formed in yields that varied from ~50% to ~80%. Several substrates underwent smooth cross-coupling with allyltrimethylsilane under comparable reaction conditions.

7
Ring-Opening Metathesis Polymerization (ROMP) and other Ring-Opening Reactions

Many of the studies concerning ring-opening metathesis by well-characterized metathesis catalysts have employed substituted norbornenes or norbornadienes. Substituted norbornenes and norbornadienes are readily available in wide variety, and they usually react irreversibly with an alkylidene. Norbornene itself is the most reactive, and the resulting polynorbornene probably is the most susceptible to secondary metathesis. Formation of polynorbornene often is used as the "test" reaction for ROMP activity. ROMP by well-defined species has been reviewed relatively recently [30], so only highlights and selected background material will be covered here.

The first report of ROMP activity by a well-characterized Mo or W species was polymerization of norbornene initiated by W(CH-t-Bu)(NAr)(O-t-Bu)$_2$ [122]. In the studies that followed, functionality tolerance, the synthesis of block copolymers, and ring-opening of other monomers were explored [30, 123]. Two important issues in ROMP concern the *cis* or *trans* nature of the double bond formed in the polymer and the polymer's tacticity. Tacticity is a consequence of the presence of two asymmetric carbons with opposite configuration in each monomer unit. The four ROMP polymers (using polynorbornene as an example) that have a regular structure are shown in Scheme 3.

A detailed study of ROMP of disubstituted norbornadienes (e.g., 2,3-dicarbomethoxynorbornadiene or 2,3-bis(trifluoromethyl)norbornadiene (NBDF6) [124] showed that they are polymerized by Mo(t-butoxide) initiators in a well-behaved living manner to give essentially monodisperse homopolymers that are

cis isotactic   cis syndiotactic

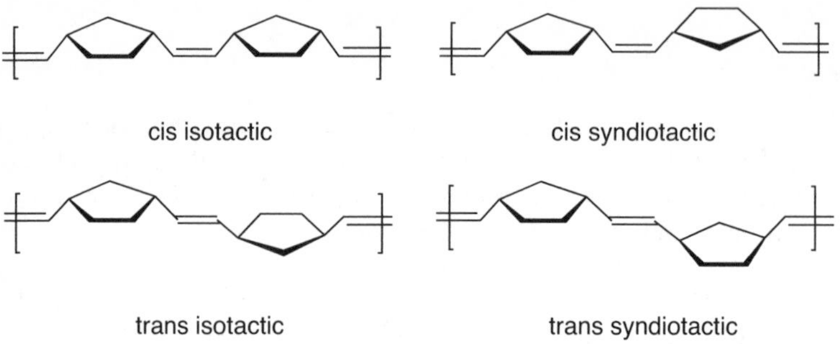

trans isotactic   trans syndiotactic

Scheme 3. The four possible regular structures of polynorbornene prepared by ring-open-ing metathesis polymerization

highly *trans* and highly tactic. Tacticity of the all *trans* polymers must be con-trolled by the chirality of the alkylidene's β carbon atom in the growing chain ("chain-end control"). In contrast, polymerizations initiated by Mo(CH-*t*-Bu)(NAr)[OCMe(CF$_3$)$_2$]$_2$ gave low polydispersity *cis*-poly(NBDF6) and *cis*-po-ly(dicarbomethoxynorbornene) that was only ~75% tactic [125]. The living na-ture of the reaction involving a relatively reactive catalyst in part can be ascribed to the relatively low reactivity of NBDF6 in general (powerful electron-with-drawing trifluoromethyl groups deactivate the olefinic bond), the low reactivity of the double bonds in the polymer (for both steric and similar electronic rea-sons), the relatively low reactivity of the "deactivated" propagating alkylidene, and the fact that propagation is believed to occur primarily via the less reactive *syn* rotamer (see below). The formation of all *trans* polymers employing the *t*-butoxide initiator and all *cis* polymers employing the hexafluoro-*t*-butoxide in-itiator is a dramatic illustration of how the nature of the alkoxide can regulate polymer structure.

A theory as to how *cis/trans* selectivity arises was the result of a series of low temperature NMR studies [77]. NBDF6 was shown to react rapidly and selective-ly with *anti*-Mo(NAr)(CHCMe$_2$Ph)[OCMe(CF$_3$)$_2$]$_2$ at –78°C in a mixture of *anti* and *syn* rotamers (generated photochemically at low temperature) to give a *syn* first-insertion product that contains a *trans* C=C bond (*anti→syn*+1$_t$; Eq. 20; P= polymer chain). At higher temperatures (up to 25°C) the *syn* rotamer reacts to produce a *syn* first-insertion product that contains a *cis* C=C bond (*syn-→syn*+1$_c$; Eq. 21). Since little *anti* form is present under normal circumstances (no photolysis) and *syn* to *anti* conversion is slow (~10^{-5} s^{-1}), *cis* polymers are proposed to be formed from *syn* species via olefin attack (through the exo face) on the CNO face of the *syn* rotamer of the catalyst with C$_7$ of the monomer ex-tending over the arylimido ring, as shown in Eq. 20.

(20)

anti   **syn+1**$_t$

(21)

syn   **syn+1**$_c$

If the mode of attack is the same in the *t*-butoxide catalyst system, where *syn* and *anti* rotamers interconvert rapidly (~1 s^{-1}), then it is possible that the mechanism for forming *trans* polymers *involves only the anti form of the propagating alkylidene species*. In short, high-*cis* polymers can be formed via *syn* intermediates when rotamer isomerization rates are negligible on the time scale of polymerization, while high-*trans* polymers can be formed via *anti* intermediates when rotamer isomerization rates are fast on the time scale of polymerization. These studies suggest that (a) in any catalyst system of this general type *syn* and *anti* rotamers (essentially two types of catalysts) might be accessible, either via rotation of the alkylidene about the Mo=CHR bond, or via reaction of the Mo=CHR bond with a C=C double bond in the substrate (i.e., as part of chain growth itself); (b) *syn* and *anti* rotamers may or may not interconvert readily on the time scale of polymerization; and (c) reactivities of *syn* and *anti* rotamers might differ by many orders of magnitude. Unfortunately, the reactivity difference between *anti* and *syn* rotamers could be confirmed only for OCMe(CF$_3$)$_2$ catalysts, since *syn* and *anti* rotamers interconvert too readily in the *t*-butoxide system. If *trans* polymer always arises via CNO face attack on an *anti* rotamer, and *cis* polymer always arises via CNO face attack on a *syn* rotamer, regardless of the type of alkoxide present, then k_a must be greater than $10^5 k_s$ in order for all *trans* polymer to result in the *t*-butoxide catalyst system (if we require $k_a[anti]$ >$10^2 k_s[syn]$ and $K_{eq}=10^3=[syn]/[anti]$). In the OCMe(CF$_3$)$_2$ system the *anti* rotamer is still much more reactive than the *syn* rotamer, but the *syn* rotamer itself in this case is also reactive.

The *cis*/*trans* ratio of polymers prepared via ROMP of 2,3-bis-(trifluorome-thyl)norbornadiene) (NBDF6) and several dicarboalkoxynorbornadienes using well-defined initiators of the type $Mo(NAr)(CHCMe_2Ph)(OR)_2$ (OR=a variety of alkoxides, including mono- and bidentate phenoxides) was found to be temper-ature dependent; at low temperatures ($-35°C$) all *cis* polymer is formed and at high temperatures ($65°C$) up to 90% *trans* polymers are formed [126]. These re-sults are consistent with formation of *cis* polymer from *syn* alkylidene rotamers and *trans* polymer from *anti* alkylidene rotamers in circumstances where *syn* and *anti* rotamers can interconvert. Base-free *syn* rotamers were found to be the species present at room temperature in the case of several phenoxide or biphe-noxide catalysts, although at low temperatures THF adducts of both *syn* and *anti* rotamers formed readily. The formation of more *trans* polymer in the phenoxide systems in general is consistent with greater accessibility to the *anti* rotamer as a consequence of more facile conversion of *syn* to *anti* rotamers.

Diolate ligands are especially attractive ligands for metathesis catalysts of the type we are discussing here in that they can be chiral and therefore could give rise to polymers whose tacticity is regulated by enantiomorphic site control. Indeed, poly(NBDF6) prepared using $(\pm)$-$Mo(N$-$2,6$-$C_6H_3Me_2)(CHCMe_2Ph)$ $[BINO(SiMe_2Ph)_2]$ $(BINO(SiMe_2Ph)_2$ is the binaphtholate that is substituted at the 3 and 3' positions with an $SiMe_2Ph$ group; see Scheme 1) as the initiator was not only >99% *cis* but was >99% tactic. Analogous all *cis*, highly tactic polymers prepared from enantiomerically pure 2,3-dicarboalkoxynorbornadienes were shown to be isotactic by proton/proton correlation spectroscopy and decoupling experiments, while the all *trans*, highly tactic polymers prepared using $Mo(CHCMe_2Ph)(NAr)(O$-t-$Bu)_2$ as the initiator were shown to be syndiotactic [127]. Related experiments employing enantiomerically pure disubstituted nor-bornenes (2,3-dicarbomethoxynorborn-5-ene, 2,3-dimethoxymethylnorborn-5-ene, and 5,6-dimethylnorborn-2-ene) gave high *cis*, isotactic or high *trans*, atactic polymers, respectively [127].

According to the model developed from reactivity studies of hexafluoro-*t*-bu-toxide complexes, *cis*, isotactic polymer should be the product of addition of mon-omer to the same CNO face of a *syn* alkylidene to give an insertion product that is a *syn* rotamer. However, at this stage it is not known whether this model holds for the binaphtholate complexes, in which interconversion of *syn* and *anti* rotamers appears to be relatively facile (as is true of phenoxide complexes in general [67]). In any case, the enormous importance of steric factors and relative rates of reac-tivities of rotamers is illustrated by the finding that the polyNBDF6 prepared using $(\pm)$-$Mo(N$-$2,6$-i-$Pr_2C_6H_3)(CHCMe_2Ph)[BINO(SiMe_2Ph)_2]$ (instead of $(\pm)$-$Mo(N$-$2,6$-$Me_2C_6H_3)(CHCMe_2Ph)[BINO(SiMe_2Ph)_2])$ as the initiator was only ~70% *cis*. The favored explanation is that the *syn* rotamer is more reactive in the N-2,6-$Me_2C_6H_3$ system than in the N-2,6-i-$Pr_2C_6H_3$ system, since less steric hindrance develops between the substituent in the *syn* alkylidene and an ortho methyl group in the N-2,6-$Me_2C_6H_3$ ligand when the substrate adds to the CNO face. The dra-matic difference in reactivity for catalysts that contain different ortho substituents in the imido ligand almost certainly will continue to be an important issue.

In a recent paper concerning the molecular weight distribution in the living polymerization of norbornene in the presence of chain-transfer agents [128], the authors were able to extract rate constants for initiation by $Mo(CHCMe_2Ph)$ $(N-2,6-i-Pr_2C_6H_3)(O-t-Bu)_2$ (k_i=0.57 $M^{-1}s^{-1}$), propagation by living poly(norbornene) (k_p=17 $M^{-1}s^{-1}$), and chain transfer by neohexene (k_{tr}=3×10^{-5} $M^{-1}s^{-1}$). Propagation is approximately an order of magnitude faster than initiation for steric reasons, while the relatively slow rate of chain transfer relative to propagation can be accounted for on the basis of the high reactivity of norbornene. Since both *syn* and *anti* rotamers are accessible in the t-butoxide system, it is uncertain to which rotamers these rate constants refer. The mixture of *cis* and *trans* double bonds in the resulting poly(norbornene) could be taken as evidence that both rotamers of the living polymer are involved.

Molybdenum catalysts have been employed to prepare a variety of poly(norbornene) or poly(norbornadiene) ROMP polymers [30]. Since 1993 papers have appeared that report polymers that contain redox-active groups [129, 130], photo- or electroluminescent groups [131–133], side-chain liquid crystals [134–140], metal nanoclusters [141–144], semiconductor nanoclusters [145–148], sugars [149], and amino acid residues [150]. Amphiphilic architectures formed via ROMP of macromonomers [151, 152] have also been reported. Studies of the polymerization of 1,7,7-trimethylbicyclo[2.2.1]hept-2-ene by $Mo(CHCMe_2Ph)$ $(NAr)(OCMe(CF_3)_2)_2$ suggest that conversion of a *syn* to an *anti* rotamer is the *rate limiting step* in this very slow polymerization reaction [125, 153–155]; only the *anti* rotamer is reactive enough toward this very bulky norbornene. This important finding suggests that the steric bulk of a monomer will be a factor in determining which rotamer will be involved in most metathesis steps. A variety of other ROMP studies with molybdenum or tungsten imido alkylidene initiators have been reported [156–159].

Both molybdenum and tungsten imido alkylidene complexes have been employed for ROMP of various cyclobutenes, including cyclobutene itself [59]. Pure linear polyethylene has been prepared by hydrogenating poly(cyclobutene) [160], "perfect" rubber has been prepared by ring-opening of 1-methylcyclobutene by $MoF6$ [161], and alternating copolymers have been prepared from 3-methylcyclobutene and 3,3-dimethylcyclobutene [162]. Polybutadienes have also been prepared via the living ROMP of 3,4-disubstituted cyclobutenes [163, 164].

The aryloxo(chloro)neopentylidenetungsten complex developed by Basset and coworkers (Scheme 2) has also been employed as an initiator in ROMP reactions [165, 166].

A combination of ring-opening and cross-coupling has been reported recently by Blechert and coworkers [167, 168]. The preference for a Mo catalyst or a Ru catalyst varied from substrate to substrate. One example where $MoF6$ was the only successful catalyst, probably because of strong binding of the cyano group to Ru, is shown in Eq. 22. Ring-opening/cross-coupling has also been reported by Basset and coworkers [169] for some sulfur-containing monomers. An example is shown in Eq. 23. The formation of unsymmetrical ring-opened/cross-coupled products in good yield would seem to have high potential in organic synthesis, especially if it could be accomplished asymmetrically.

$$(22)$$

$$(23)$$

8
Conjugated Polymers

The synthesis of conjugated polymers such as polyacetylenes, paraphenylenevinylene (PPV), etc., some in a living manner, has enjoyed a renaissance now that well-defined initiators are available. Soluble, alkyl-substituted polyacetylene can be prepared from monosubstituted cyclooctatetraenes using tungsten imido catalysts [170], an approach that was initiated by Grubbs and coworkers several years ago [30]. Polyenes also can be prepared indirectly by polymerization of a norbornene followed by thermolysis [156, 171–175]. Bulky acetylenes such as ortho-trimethylsilylphenylacetylene can be polymerized in a living manner by catalysts that contain a "small" alkoxide such as hexafluoroisopropoxide if a base is bound to the metal that has the correct lability [37, 38]. (It had been known for some time that such species can be polymerized by olefin metathesis catalysts prepared by classical techniques [32, 33, 176].) Finally, polyenes can be prepared in a living manner from 1,6-heptadiyne derivatives [35, 36, 177, 178].

Metals have been incorporated into polyenes, either in the side chain as a consequence of the living polymerization of ethynylferrocene and ethynylruthenocene [179], or in the main chain as a consequence of polymerization of poly(ferrocenylenedivinylene) and poly(ferrocenylenebutenylene) [180]. The former polymers were prepared in a living manner; polymers containing as many as 40 equivalents of monomer were still soluble. In the latter only relatively low molecular weight materials were obtained.

Poly(1,4-naphthylenevinylenes) have been prepared by metathesis polymerization of benzobarrelenes [181, 182] and the photoluminescence properties of homopolymers and block-copolymers have been studied in some detail [183]. PPV also has been prepared via ROMP of [2.2]paracyclophane-1,9-diene [184] and ROMP of a paracyclophene that contains a solubilizing leaving group [185]. The resulting polymer is converted to PPV upon acid catalysis at room temperature. ADMET of 2,5-dialkyl-1,4-divinylbenzenes using Mo or W catalysts has been a successful route to soluble PPV [186–188]. The most desirable catalyst (in terms of speed and selectivity) was found to be $Mo(CHCMe_2Ph)(N-2,6-Me_2C_6H_3)[OCMe(CF_3)_2]_2$ [186]. Interestingly, resonances in the proton NMR spectrum at –0.3 and –1.2 ppm provided evidence that a molybdenacyclobutane complex is present in equilibrium with the expected alkylidene complexes during the reaction. Molybdenacyclobutane complexes have not been observed during a metathesis reaction before.

9
Ring-Closing Metathesis (RCM)

Ring-closing metathesis of diolefins has been a recognized variant of olefin metathesis for more than 15 years [189–191]. Although RCM (as it is now known) had not been used extensively in organic synthesis, possibly in view of the fact that only classical catalysts could be employed [1], examples such as the synthesis of 3-pentenecarboxylic acid esters using WCl_6 activated with silanes still appear today [192]. However, low yields, irreproducibility, and intolerance of functionalities tend to be common when classical catalysts are employed. With the advent of well-defined catalysts the reproducibility and yields have improved markedly and functionalities are now tolerated to an ever increasing extent. The two main types of catalysts that are used for RCM currently are molybdenum imido catalysts (usually $Mo(CHMe_2Ph)(N-2,6-i-Pr_2C_6H_3)[OCMe(CF_3)_2]_2$, or "MoF6") or ruthenium catalysts of the type $Ru(CHR)Cl_2(PCy_3)_2$ (e.g., CHR= CHPh or $CHCH=CPh_2$). The ruthenium catalysts have the advantage of a greater tolerance of functionality, but sacrifice speed. The molybdenum catalyst has the advantage of being able to form rings in sterically more demanding circumstances, but is relatively intolerant of protons on heteroatoms (carboxylic acids, alcohols, thiols, etc.) and some functionalities (e.g., aldehydes). A recent comparison of ADMET catalyzed by MoF6 and $Ru(CHPh)Cl_2(PCy_3)_2$ [28] reveals that Mo is more reactive by a factor of ~20 toward 1,9-decadiene, but relatively much more reactive when O or S is present. For example, the rate of ethylene evolution in ADMET of $[CH_2=CH(CH_2)_3S]_2$ is lower by a factor of 10 compared to $[CH_2=CH(CH_2)_3]_2$ when Mo is employed, but is essentially zero when Ru is employed. It is surprising that the rate of metathesis by Ru is so dramatically attenuated in the presence of donors such as O, S, CN, or amines, while the rate of metathesis by MoF6 is much less so. A casual analysis would have suggested that the opposite would be true.

Both Mo and Ru catalysts have begun to show their utility in organic synthesis involving RCM. Molybdenum catalysts come in many variations and covalently bound ligands can be architecturally designed to achieve a desired steric goal. However, most of the work so far has employed the MoF6 catalyst. Aryloxide alkylidene complexes of tungsten that do not contain any oxo or imido ligand in the initial species have also been employed [106–108, 165, 166, 169, 193–195] for RCM, as have systems that consist of $W(O)(OAryl)_2Cl_2$ activated by lead alkyls [45, 196]. Only the RCM reactions catalyzed by Mo or W catalysts will be discussed here. Applications of ring-closing metathesis to organic synthesis have been discussed in recent articles [10, 197–199] and appear elsewhere in this volume. Cycloalkenes have also been synthesized via alkylidene-mediated olefin metathesis and carbonyl olefination [200], but this approach is relatively unattractive for ring-closing on any significant scale because of its stoichiometric nature. However, the method could be viable in circumstances where the alkylidene complex is relatively plentiful and/or the product is rare and especially valuable [201].

Five-membered carbocycles are the most easily formed [45, 107, 196, 200, 202, 203]. Five-membered carbocyclic rings can be formed (with 2% MoF6 as the catalyst) even when the double bond is tetrasubstituted (Eq. 24) [200]. The stability of the catalyst toward the free OH group in this case is noteworthy. Evidently this particular "t-butoxide-like" alcohol does not react with this particular catalyst for steric reasons. Six-membered carbocyclic rings are also formed readily (Eq. 25) [200], as are seven-membered rings, especially if one takes advantage of a Thorpe-Ingold effect (e.g., Eq. 26) [20] or a similar conformational predisposition for the double bonds to remain near one another.

$$(24)$$

$$(25)$$

$$(26)$$

A recent paper [204] compares the ring-closing ability of Ru and MoF6 catalysts for gem-disubstituted olefins to five-, six-, and seven-membered tri- and tetrasubstituted carbocycles (primarily). The findings demonstrate that ring-closing to give more sterically hindered cyclic olefins is much slower with Ru than Mo, although Ru gives good yields of some trisubstituted cyclic olefins. The time required for Ru was as long as 48 h, but the minimum times required for Mo (at 65°C) were not determined. Since comparable side-by-side rate studies were not carried out, it is difficult to say on an absolute scale what are the relative rates of Ru and Mo in a given circumstance. It should be noted that the favored *syn* rotamer of the MoF6 catalyst converts to the *anti* rotamer with a rate constant of approximately 10^{-5} s^{-1}, and that the *anti* rotamer was found to be as much as five orders of magnitude more reactive than the *syn* rotamer in one circumstance [77]. Therefore a temperature of 65°C may be needed for some reactions involving MoF6 in order that the rate constant for conversion of *syn* to *anti* rotamers might be as high at 10^{-3} at that temperature and the half-life for conversion of a *syn* to an *anti* rotamer therefore reduced from hours (at 22°C) to minutes (at 65°C). (See also the studies concerned with polymerization of 1,7,7-trimethylbicyclo[2.2.1]hept-2-ene [125, 153–155].) These studies demonstrate that ruthenium catalysts are active enough to produce cyclic species containing up to trisubstituted double bonds in some five-, six-, or even seven-membered rings, and suggest that Mo and Ru catalysts often will play different but important roles in the construction of complex organic molecules through RCM.

Rings that contain ether linkages have been prepared (e.g., Eqs. 27 [20], 28 [205], or 29 [205]). Even five-membered cyclic ethers that contain trisubstituted double bonds can be formed using MoF6 as the catalyst [206]. Ru(CHCH= CPh$_2$)Cl$_2$(PCy$_3$)$_2$ did not catalyze metathesis of a variety of acyclic enol ethers that were reported to be ring-closed by MoF6 [205].

(27)

(28)

(29)

The formation of rings that contain a thioether linkage does not appear to be catalyzed efficiently by Ru, even when terminal olefins are present. On the other hand, molybdenum appears to work relatively well, as shown in Eqs. 30 [207] and 31 [208]. Under some conditions polymerization (ADMET) to give polythioethers is a possible alternative [26]. Aryloxide tungsten catalysts have also been employed successfully to prepare thioether derivatives [107, 166, 169]. Apparently the mismatch between a "hard" earlier metal center and a "soft" sulfur donor is what allows thioethers to be tolerated by molybdenum and tungsten. Similar arguments could be used to explain why cyclometalated aryloxycarbene complexes of tungsten have been successfully employed to prepare a variety of cyclic olefins such as the phosphine shown in Eq. 32 [107, 193].

$$\text{(30)}$$

$$\text{(31)}$$

$$\text{(32)}$$

A variety of cyclic olefins (5-, 6-, and 7-membered) that contain nitrogen have been prepared via ring-closing metathesis, for example as shown in Eq. 33 [209]. Other examples are shown in Eqs. 34 [210] and 35 [211]. A variety of pyrrolizidines, indolizidines, quinolizidines, pyrrolidinoazocines, piperidinoazocines, and other fused nitrogen heterocycles have also been prepared via RCM (e.g., Eq. 36 [212, 213]).

$$\text{(33)}$$

$$\text{(34)}$$

$$\text{(35)}$$

n = 1, 2, 3   n = 1 (84%), 2 (53%), 3 (12%)

$$(36)$$

Tungsten aryloxo complexes have been shown to catalyze the intramolecular metathesis reactions of di- and tri-substituted ω-unsaturated glucose and glucosamine derivatives to yield bicyclic carbohydrate-based compounds containing 12- and 14-membered rings [108, 214, 215]. An example is shown in Eq. 37. The tolerance for amides and esters is noteworthy, as are the yields and the size of the rings that are formed.

$$(37)$$

Hoveyda and coworkers [216] have shown that chromenes can be prepared via the ring-opening/closing sequence shown in Eq. 38, catalyzed by either $Ru(CH-CH=CPh_2)Cl_2(PCy_3)_2$ or MoF6. The MoF6 catalyst consistently gave higher yields (often 90–95%) under the conditions employed. For example, the cyclohexenyl substrate shown gave a 90% yield of the product with Mo, but only 32% with Ru.

$$(38)$$

RCM has been employed to prepare relatively large rings and natural products. For example, the enantiomerically pure tetracyclic ABCE subunit of manzamine A has been constructed by a series of reactions that is completed by a ring-closing metathesis to form the azocine ring [217]. A trisubstituted olefin in a *trans*-fused cyclohexene ring has been formed by Mo-catalyzed RCM (Eq. 39) as part of a synthesis of coronafacic acid [218], while the eight-membered ring of dactylol (Eq. 40) was formed in high yield similarly [219].

$$(39)$$

$$(40)$$

92%

$$(41)$$

90%

The analogous *cis-* fused [3.6]bicycle was also prepared readily. Hoveyda and coworkers have used a Mo-catalyzed RCM step in the enantioselective total synthesis of antifungal agent SCH-38516 (Eq. 41) [220, 221]. In a recent full paper [46] it was discovered that freshly prepared or recrystallized Mo catalyst allowed ring-closing metathesis to occur smoothly at 22°C to afford the cyclic product shown in Eq. 41 in 90% yield after only 4 h. Only dimer was formed with Ru catalysts. This result illustrates that large rings can be constructed by RCM, and raises the possibility that coordination of various donor functionalities to the metal during the cyclization process could be beneficial.

The six or seven-membered cyclic ethers that comprise the Brevetoxin subunits could be prepared by Mo-catalyzed RCM (Eq. 42) [13]. It is noteworthy that the most successful substrates were those in which R_1=H and R_2=Me. The first metal-containing intermediate in that circumstance therefore is likely to be that in which there is an oxygen bound to the alkylidene carbon atom; yet that alkylidene behaves in a ring-closing reaction in the expected manner.

$$(42)$$

$$(43)$$

$$\text{(44)}$$

a $R_1 = H$, $R_2 = Et$
b $R_1 = Et$, $R_2 = H$
$R_3 = $ p-MeOC$_6$H$_4$
$n = 1$ or 2

High oxidation state alkylidene complexes in which a heteroatom is bound to the alkylidene carbon atom are extremely rare [41]. Since the approach shown in Eq. 43 failed, the related approach shown in Eq. 44 was taken to prepare the medium-sized ring subunits [222]. The latter product was formed in good yield when n=2, R_1=H, R_2=Et, but only poor yield when n=2, R_1=Et, R_2=H, possibly "due to unfavorable interactions between the ethyl substituent and transannular groups in the transition state for cyclization of the allyl ether" [222]. Ruthenium catalysts either failed or gave low yields, presumably because of the steric hindrance associated with ring-closing dienes of this type.

10
Asymmetric RCM

Molybdenum catalysts that contain enantiomerically pure diolates are prime targets for asymmetric RCM (ARCM). Enantiomerically pure molybdenum catalysts have been prepared that contain a tartrate-based diolate [86], a binaphtholate [87], or a diolate derived from a *trans*-1,2-disubstituted cyclopentane [89, 90], as mentioned in an earlier section. A catalyst that contains the diolate derived from a *trans*-1,2-disubstituted cyclopentane has been employed in an attempt to form cyclic alkenes asymmetrically via kinetic resolution (inter alia) of substrates **A** and **B** (Eqs. 45, 46) where OR is acetate or a siloxide [89, 90]. Reactions taken to ~50% consumption yielded unreacted substrate that had an ee between 20% and 40%. When **A** (OR=acetate) was taken to 90% conversion, the ee of residual **A** was 84%. The relatively low enantioselectivity might be ascribed to the slow interconversion of *syn* and *anti* rotamers of the intermediates or to the relatively "floppy" nature of the diolate that forms a pseudo nine-membered ring containing the metal.

$$\text{(45)}$$

$$\text{(46)}$$

A Mo catalyst that contains an enantiomerically pure binaphtholate or biphenolate ligand would seem to have a better opportunity for ARCM since the ligand forms a relatively rigid pseudo six-membered ring containing the metal and *syn* and *anti* rotamers are known to interconvert readily. Recently an enantiomerically pure $Mo(CHCMe_2Ph)(N-2,6-i-Pr_2C_6H_3)(O_2R^*)$ complex (O_2R^*=5,5',6,6'-tetramethyl-3,3'-di-*tert*-butyl-1,1'-biphenyl-2,2'-diolate) has been prepared and employed for ARCM of substrates of the type shown in Eqs. 45 and 46 [223]. Conversion of one enantiomer of **A** (Eq. 45; R=SiEt₃) in a racemic mixture to give the ring-closed product in ~50% yield and ~95% ee suggests that this type of catalyst, or variations of it, are likely to have wide application in the construction of rings in an asymmetric manner.

11
Comments and Conclusions

It is clear that "well-defined" Mo and W metathesis catalysts have tremendous advantages over classical metathesis catalysts for the synthesis of polymers by ROMP or ADMET, or variations of such reactions. They also are proving to be useful for the synthesis of cyclic organic compounds, since such syntheses usually are dramatically shorter than syntheses employing traditional organic methods. In the near future new catalyst variations will be developed and we will understand in greater depth their mode of operation at a molecular level. Turnover frequencies are likely to increase, perhaps by more than an order of magnitude, as we become aware of the factors that limit a catalyst's lifetime and selectivity. The development of enantiopure catalysts for asymmetric metathesis reactions shows great promise, although almost certainly no "magic catalyst" will accomplish all asymmetric reactions with equal ease; therefore catalyst design will play an ever more significant role in such reactions. Although Ru catalysts have the advantage of greater tolerance of water, oxygen, and protic reagents, their reactivity appears to be significantly lower than reactivities of the more sensitive Mo and W catalysts, especially in the presence of good donors such as S or P, and even O. With time a collection of Mo, W, Ru, and perhaps yet other catalysts will emerge whose advantages and disadvantages we will learn to know and manipulate. Finally, although acetylene metathesis by "well-defined" alkylidyne catalysts [60] has been known for over 15 years, and some of the principles are closely related to those of olefin metathesis, acetylene metathesis has not yet entered the picture in terms of the synthesis of organic compounds. Therefore, acetylene metathesis would seem to be ripe for development, in particular for synthesis of large rings containing C≡C bonds. Eventually a wide variety of catalyst variations will be available for testing a wide variety of substrates and reactions. The amount of work involved in these endeavors will be large, but the payoffs appear to be worth that effort.

Acknowledgements: R.R.S. thanks the National Science Foundation for supporting fundamental studies of complexes containing multiple metal–carbon bonds, the Navy for supporting studies in ring-opening metathesis polymerization, the Department of Energy for supporting studies in the synthesis of polyenes, and the graduate and postdoctoral students who have contributed their time, energy, and talent.

12
References

1. Ivin KJ, Mol JC (1997) Olefin metathesis and metathesis polymerization. Academic Press, San Diego
2. Grubbs RH, Tumas W (1989) Science 243:907
3. Gilliom LR, Grubbs RH (1986) J Am Chem Soc 108:733
4. Gilliom LR, Grubbs RH (1986) Organometallics 5:721
5. Schrock RR (1986) In: Braterman PR (ed) Reactions of coordinated ligands. Plenum, New York
6. Schrock RR (1986) J Organometal Chem 300:249
7. Schrock RR (1995) Polyhedron 14:3177
8. Schwab P, France MB, Ziller JW, Grubbs RH (1995) Angew Chem Int Ed Engl 34:2039
9. Schwab P, Grubbs RH, Ziller JW (1996) J Am Chem Soc 118:100
10. Grubbs RH, Miller SJ, Fu GC (1995) Acc Chem Res 28:446
11. Demonceau A, Stumpf AW, Saive E, Noels AF (1997) Macromolecules 30:3127
12. Thorn-Csányi E, Dehmel J, Luginsland HD, Zilles JU (1997) J Mol Catal A 115:29
13. Clark JS, Kettle JG (1997) Tet Lett 38:123
14. Fox HH, Schrock RR (1994) Organometallics 13:635
15. Konzelman J, Wagener KB (1996) Macromolecules 29:7657
16. Konzelman J, Wagener KB (1995) Macromolecules 28:4686
17. Marmo JC, Wagener KB (1995) Macromolecules 28:2602
18. Wagener KB, Patton JT, Forbes MDE, Myers TL, Maynard HD (1993) Poly Int 32:411
19. Marmo JC, Wagener KB (1993) Macromolecules 26:2137
20. Forbes MDE, Patton JT, Myers TL, Smith Jr. DW, Schulz GR, Wagener KB (1992) J Am Chem Soc 114:10978
21. Brzezinska K, Wagener KB (1992) Macromolecules 25:2049
22. Wagener KB, Patton JT (1993) Macromolecules 26:249
23. Smith DW, Wagener KB (1993) Macromolecules 26:3533
24. Smith DW, Jr., Wagener KB (1993) Macromolecules 26:1633
25. Ogara JE, Wagener KB (1993) Makromol Chem – Rapid Comm 14:657
26. O'Gara JE, Portmess JD, Wagener KB (1993) Macromolecules 26:2837
27. Patton JT, Boncella JM, Wagener KB (1992) Macromolecules 25:3862
28. Wagener KB, Brzezinska K, Anderson JD, Younkin TR, Steppe K, DeBoer W (1997) Macromolecules 30:7363
29. Wolfe PS, Gomez FJ, Wagener KB (1997) Macromolecules 30:714
30. Schrock RR (1993) In: Brunelle DJ (ed) Ring-opening polymerization. Hanser, Munich
31. Masuda T (1997) In: Kobayashi S (ed) Catalysis in precision polymerization. John Wiley & Sons, New York
32. Masuda T, Higashimura T (1984) Acc Chem Res 17:51
33. Masuda T, Higashimura T (1987) Adv Polym Sci 81:122
34. Shirakawa H, Masuda T, Takeda K (1994) In: Patai S (ed) Supplement C2: the chemistry of triple-bonded functional groups. John Wiley & Sons, New York
35. Schattenmann FJ, Schrock RR, Davis WM (1996) J Am Chem Soc 118:3295
36. Schattenmann FJ, Schrock RR (1996) Macromolecules 29:8990
37. Schrock RR, Luo S, Lee JC, Zanetti NC, Davis WM (1996) J Am Chem Soc 118:3883
38. Schrock RR, Luo S, Zanetti N, Fox HH (1994) Organometallics 13:3396
39. Kim SH, Zuercher WJ, Bowden NB, Grubbs RH (1996) J Org Chem 61:1073

40. Peters JU, Blechert S (1997) Chem Commun 1983
41. Feldman J, Schrock RR (1991) Prog Inorg Chem 39:1
42. Buhro WE, Chisholm MH (1987) Adv Organometal Chem 27:311
43. Breslow DS (1993) Progress in Polymer Science 18:1141
44. Streck R (1988) J Mol Catal 46:305
45. Nugent WA, Feldman J, Calabrese JC (1995) J Am Chem Soc 117:8992
46. Xu Z, Johannes CW, Houri AF, La DS, Cogan DA, Hofilena GE, Hoveyda AH (1997) J Am Chem Soc 119:10302
47. Parkin G, Bunel E, Burger BJ, Trimmer MS, van Asselt A, Bercaw JE (1987) J Molec Catal 41:21
48. Freundlich JS, Schrock RR, Davis WM (1996) J Am Chem Soc 118:3643
49. Freundlich JS, Schrock RR, Davis WM (1996) Organometallics 15:2777
50. Edwards DS, Biondi LV, Ziller JW, Churchill MR, Schrock RR (1983) Organometallics 2:1505
51. van der Schaff PA, Hafner A, Mühlebach A (1996) Angew Chem Int Ed Engl 35:1845
52. Johnson LK, Frey M, Ulibarri TA, Virgil SC, Grubbs RH, Ziller JW (1993) J Am Chem Soc 115:8167
53. Sharp PR, Schrock RR (1979) J Organometal Chem 171:43
54. Wengrovius JH, Schrock RR (1982) Organometallics 1:148
55. Binger P, Müller P, Benn R, Mynott R (1989) Angew Chem 101:647
56. Johnson LK, Grubbs RH, Ziller JW (1993) J Am Chem Soc 115:8130
57. de la Mata FJ, Grubbs RH (1996) Organometallics 15:577
58. de la Mata FJ (1996) J Organomet Chem 525:183
59. Wu Z, Wheeler DR, Grubbs RH (1992) J Am Chem Soc 114:146
60. Murdzek JS, Schrock RR (1988) In: (ed) Carbyne complexes. VCH, New York
61. Schaverien CJ, Dewan JC, Schrock RR (1986) J Am Chem Soc 108:2771
62. Schrock RR, DePue RT, Feldman J, Yap KB, Yang DC, Davis WM, Park LY, DiMare M, Schofield M, Anhaus J, Walborsky E, Evitt E, Krüger C, Betz P (1990) Organometallics 9:2262
63. Schrock RR, DePue R, Feldman J, Schaverien CJ, Dewan JC, Liu AH (1988) J Am Chem Soc 110:1423
64. Feldman J, DePue RT, Schaverien CJ, Davis WM, Schrock RR (1989) In: Schubert U (ed) Advances in metal carbene chemistry. Kluwer, Boston
65. Schrock RR, Murdzek JS, Bazan GC, Robbins J, DiMare M, O'Regan M (1990) J Am Chem Soc 112:3875
66. Kress J, Osborn JA (1992) Angew Chem, Int Ed Engl 31:1585
67. Schrock RR, Crowe WE, Bazan GC, DiMare M, O'Regan MB, Schofield MH (1991) Organometallics 10:1832
68. Wu Y-D, Peng Z-H (1997) J Am Chem Soc 119:8043
69. Cundari TR, Gordon MS (1992) J Am Chem Soc 114:539
70. Ushio J, Nakatsuji H, Yonezawa T (1984) J Am Chem Soc 106:5892
71. Fox HH, Schofield MH, Schrock RR (1994) Organometallics 13:2804
72. Cundari TR, Gordon MS (1992) J Am Chem Soc 114:539
73. Cundari TR, Gordon MS (1992) Organometallics 11:55
74. Folga E, Ziegler T (1993) Organometallics 12:325
75. Brookhart M, Green MLH, Wong LL (1988) Prog Inorg Chem 36:1
76. Oskam JH, Schrock RR (1992) J Am Chem Soc 114:7588
77. Oskam JH, Schrock RR (1993) J Am Chem Soc 115:11831
78. Bell A, Clegg W, Dyer PW, Elsegood MRJ, Gibson VC, Marshall EL (1994) J Chem Soc, Chem Commun 2547
79. Bell A, Clegg W, Dyer PW, Elsegood MRJ, Gibson VC, Marshall EL (1994) J Chem Soc Chem Commun 2247
80. Schoettel G, Kress J, Osborn JA (1989) J Chem Soc, Chem Commun 1062
81. Fox HH, Yap KB, Robbins J, Cai S, Schrock RR (1992) Inorg Chem 31:2287

82. Fox HH, Lee J-K, Park LY, Schrock RR (1993) Organometallics 12:759
83. Oskam JH, Fox HH, Yap KB, McConville DH, O'Dell R, Lichtenstein BJ, Schrock RR (1993) J Organometal Chem 459:185
84. Buchmeiser M, Schrock RR (1995) Inorg Chem 34:3553
85. Luo, S, Schrock, RR, unpublished observations
86. McConville DH, Wolf JR, Schrock RR (1993) J Am Chem Soc 115:4413
87. Totland KM, Boyd TJ, Lavoie GG, Davis WM, Schrock RR (1996) Macromolecules 29:6114
88. Dietz SD, Eilerts NW, Heppert JA, Morgan MD (1993) Inorg Chem 32:1698
89. Fujimura O, Delamata FJ, Grubbs RH (1996) Organometallics 15:1865
90. Fujimura O, Grubbs RH (1996) J Am Chem Soc 118:2499
91. Feher FJ, Tajima TL (1994) J Am Chem Soc 116:2145
92. VanderLende DD, Abboud KA, Boncella JM (1994) Organometallics 13:3378
93. Vaughan WM, Abboud KA, Boncella JM (1995) Organometallics 14:1567
94. Vaughan WM, Abboud KA, Boncella JM (1995) J Organometal Chem 485:37
95. Vaughan WM, Abboud KA, Boncella JM (1995) J Am Chem Soc 117:11015
96. Gamble AS, Boncella JM (1993) Organometallics 12:2814
97. Blosch LL, Gamble AS, Boncella JM (1992) J Molec Catal 76:229
98. Blosch LL, Gamble AS, Abboud K, Boncella JM (1992) Organometallics 11:2342
99. Kress J, Aguero A, Osborn JA (1986) J Mol Catal 36:1
100. Aguero A, Osborn JA (1988) New J Chem 12:111
101. van der Schaaf PA, Smeets WJJ, Spek AL, van Koten G (1992) J Chem Soc, Chem Commun 717
102. van der Schaaf PA, Boersma J, Smeets WJJ, Spek AL, van Koten G (1993) Inorg Chem 32:5108
103. van der Schaaf PA, Boersma J, Kooijman H, Spek AL, van Koten G (1993) Organometallics 12:4334
104. van der Schaaf PA, Abbenhuis RATM, Grove DM, Smeets WJJ, Spek AL, van Koten G (1993) J Chem Soc, Chem Commun 504
105. van der Schaaf PA, Abbenhuis RATM, van der Noort WPA, Degraaf R, Grove DM, Smeets WJJ, Spek AL, van Koten G (1994) Organometallics 13:1433
106. Lefebvre F, Leconte M, Pagano S, Mutch A, Basset JM (1995) Polyhedron 14:3209
107. Leconte M, Pagano S, Mutch A, Lefebvre F, Basset JM (1995) Bull Soc Chim Fr 132:1069
108. Descotes G, Ramza J, Basset JM, Pagano S, Gentil E, Banoub J (1996) Tetrahedron 52:10903
109. Heppert JA, Dietz SD, Eilerts NW, Henning RW, Morton MD, Takusagawa F, Kaul FA (1993) Organometallics 12:2565
110. Keller A, Matusiak R (1996) J Mol Catal A 104:213
111. Keller A (1992) J Organomet Chem 436:199
112. Buffon R, Leconte M, Choplin A, Basset J-M (1994) J Chem Soc, Dalton Trans 1723
113. Buffon R, Leconte M, Choplin A, Basset J-M (1993) J Chem Soc, Chem Commun 361
114. Weiss K, Lössel G (1989) Angew Chem, Int Ed Engl 28:62
115. Herrmann WA, Stumpf AW, Priermeier T, Bogdanovic S, Dufaud V, Basset JM (1997) Angew Chem Int Ed 35:2803
116. Crowe WE, Zhang ZJ (1993) J Am Chem Soc 115:10998
117. Crowe WE, Goldberg DR (1995) J Am Chem Soc 117:5162
118. Crowe WE, Goldberg DR, Zhang ZJ (1996) Tetrahedron Lett 37:2117
119. Brümmer O, Rückert A, Blechert S (1997) Chem-Eur J 3:441
120. Barrett AGM, Beall JC, Gibson VC, Giles MR, Walker GLP (1996) Chem Commun 2229
121. Feng J, Schuster M, Blechert S (1997) Synlett 129
122. Schrock RR, Feldman J, Cannizzo L, Grubbs RH (1987) Macromolecules 20:1169
123. Schrock RR (1990) Acc Chem Res 23:158
124. Bazan G, Khosravi E, Schrock RR, Feast WJ, Gibson VC, O'Regan MB, Thomas JK, Davis WM (1990) J Am Chem Soc 112:8378

125. Feast WJ, Gibson VC, Marshal EL (1992) J Chem Soc, Chem Commun 1157
126. Schrock RR, Lee J-K, O'Dell R, Oskam JH (1995) Macromolecules 28:5933
127. O'Dell R, McConville DH, Hofmeister GE, Schrock RR (1994) J Am Chem Soc 116:3414
128. Benedicto AD, Claverie JP, Grubbs RH (1995) Macromolecules 28:500
129. Albagli D, Bazan G, Schrock RR, Wrighton MS (1993) J Phys Chem 97:10211
130. Albagli D, Bazan GC, Schrock RR, Wrighton MS (1993) J Am Chem Soc 115:7328
131. Lee J-K, Schrock RR, Baigent DR, Friend RH (1995) Macromolecules 28:1966
132. Baigent DR, Friend RH, Lee JK, Schrock RR (1995) Synth Met 71:2171
133. Boyd TJ, Geerts Y, Lee J-K, Fogg DE, Lavoie GG, Schrock RR, Rubner MF (1997) Macromolecules 30:3553
134. Komiya Z, Schrock RR (1993) Macromolecules 26:1387
135. Komiya Z, Schrock RR (1993) Macromolecules 26:1393
136. Saunders RS, Cohen RE, Schrock RR (1994) Acta Polymerica 45:301
137. Ungerank M, Winkler B, Eder E, Stelzer F (1997) Macromol Chem Phys 198:1391
138. Pugh C, Dharia J, Arehart SV (1997) Macromolecules 30:4520
139. Pugh C, Kiste AL (1997) Prog Polym Sci 22:601
140. Winkler B, Rehab A, Ungerank M, Stelzer F (1997) Macromol Chem Physics 198:1417
141. Ng Cheong Chan Y, Craig GSW, Schrock RR, Cohen RE (1992) Chem Mater 4:885
142. Ng Cheong Chan Y, Schrock RR, Cohen RE (1992) Chem Mater 4:24
143. Ng Cheong Chan Y, Schrock RR, Cohen RE (1992) J Am Chem Soc 114:7295
144. Ng Cheong Chan Y, Schrock RR, Cohen RE (1993) Chem Mater 5:566
145. Fogg DE, Radzilowski LH, Blanski R, Schrock RR, Thomas EL (1997) Macromolecules 30:417
146. Sankaran V, Yue J, Cohen RE, Schrock RR, Silbey RJ (1993) Chem Mater 5:1133
147. Tassoni R, Schrock RR (1994) Chem Mater 6:744
148. Yue J, Sankaran V, Cohen RE, Schrock RR (1993) J Am Chem Soc 115:4409
149. Nomura K, Schrock RR (1996) Macromolecules 29:540
150. Coles MP, Gibson VC, Mazzariol L, North M, Teasdale WG, Williams Cm, Zamuner D (1994) J Chem Soc, Chem Commun 2505
151. Heroguez V, Gnanou Y, Fontanille M (1997) Macromolecules 30:4791
152. Feast WJ, Gibson VC, Johnson AF, Khosravi E, Mohsin MA (1997) J Mol Catal A 115:37
153. Feast WJ, Gibson VC, Ivin KJ, Kenwright AM, Khosravi E (1994) J Mol Catal 90:87
154. Feast WJ, Gibson VC, Ivin KJ, Kenwright AM, Khosravi E (1994) J Chem Soc, Chem Commun 2737
155. Feast WJ, Gibson VC, Ivin KJ, Kenwright AM, Khosravi E (1994) J Chem Soc, Chem Commun 1399
156. Dounis P, Feast WJ, Kenwright AM (1995) Polymer 36:2787
157. Hillmyer MA, Grubbs RH (1993) Macromolecules 26:872
158. Hillmyer MA, Grubbs RH (1995) Macromolecules 28:8662
159. Schitter RME, Steinhausler T, Stelzer F (1997) J Mol Catal A 115:11
160. Wu Z, Grubbs RH (1994) Macromolecules 27:6700
161. Wu Z, Grubbs RH (1994) J Mol Catal 90:39
162. Wu Z, Grubbs RH (1995) Macromolecules 28:3502
163. Perrott MG, Novak BM (1996) Macromolecules 29:1817
164. Perrott MG, Novak BM (1995) Macromolecules 28:3492
165. Couturier JL, Leconte M, Basset JM, Ollivier J (1993) J Organomet Chem 451:C7
166. Couturier JL, Tanaka K, Leconte M, Basset JM (1993) Phosphor Sulfur Silicon 74:383
167. Schneider MF, Blechert S (1996) Angew Chem Int Ed Engl 35:411
168. Schneider MF, Lucas N, Velder J, Blechert S (1997) Angew Chem Int Ed Engl 36:257
169. Couturier JL, Tanaka K, Leconte M, Basset JM, Ollivier J (1993) Angew Chem Int Ed Engl 32:112
170. Gorman CB, Ginsburg EJ, Grubbs RH (1993) J Am Chem Soc 115:1397
171. Craig GSW, Cohen RE, Schrock RR, Esser A, Schrof W (1995) Macromolecules 28:2512

172. Craig GSW, Cohen RE, Schrock RR, Silbey RJ, Puccetti G, Ledoux I, Zyss J (1993) J Am Chem Soc 115:860
173. Craig GSW, Cohen RE, Schrock RR, Dhenaut C, LeDoux I, Zyss J (1994) Macromolecules 27:1875
174. Samuel IDW, Ledoux I, Dhenaut C, Zyss J, Fox H, Schrock RR, Silbey RJ (1994) Science 265:1070
175. Dounis P, Feast WJ, Widawski G (1997) J Molec Catal A 115:51
176. Masuda T, Fujimori JI, Abrahman MZ, Higashimura T (1993) Polymer J 25:535
177. Fox HH, Schrock RS (1992) Organometallics 11:2763
178. Fox HH, Wolf MO, O'Dell R, Lin BL, Schrock RR, Wrighton MS (1994) J Am Chem Soc 116:2827
179. Buchmeiser M, Schrock RR (1995) Macromolecules 28:6642
180. Stanton CE, Lee TR, Grubbs RH, Lewis NS, Pudelski JK, Callstrom MR, Erickson MS, McLaughlin ML (1995) Macromolecules 28:8713
181. Wagaman MW, Grubbs RH (1997) Macromolecules 30:3978
182. Pu L, Wagaman MW, Grubbs RH (1996) Macromolecules 29:1138
183. Wagaman MW, Bellmann E, Grubbs RH (1997) Phil Trans Roy Soc London A 355:727
184. Thorn-Csányi E, Hohnk HD, Pflug KP (1993) J Mol Catal 84:253
185. Miao Y-J, Bazan GC (1994) J Am Chem Soc 116:9379
186. Thorn-Csányi E, Kraxner P (1997) J Mol Catal A 115:21
187. Thorn-Csányi E, Kraxner P (1995) Macromol Chem, Rapid Commun 16:147
188. Thorn-Csányi E, Pflug KP (1993) Macromol Chem, Rapid Commun 14:619
189. Villemin D (1980) Tet Lett 21:1715
190. Tsuji J, Hashiguchi S (1980) Tet Lett 21:2955
191. Plugge MFC, Mol JC (1991) Synlett 507
192. Bespalova NB, Bovina MA, Sergeeva MB, Zailin VG (1994) Russ Chem Bull 43:1425
193. Leconte M, Jourdan I, Pagano S, Lefebvre F, Basset JM (1995) J Chem Soc Chem Commun 857
194. Couturier JL, Leconte M, Basset JM (1993) Transition Metal Carbyne Com 392:39
195. Leconte M, Basset JM, Quignard F, Larroche C (1986) In: Braterman PS (ed) Reactions of coordinated ligands. Plenum, New York
196. Martinez LE, Nugent WA, Jacobsen EN (1996) J Org Chem 61:7963
197. Schmalz HG (1995) Angew Chem Int Ed Engl 34:1833
198. Fürstner A (1977) Top Catal 4:285
199. Schuster M, Blechert S (1997) Angew Chem Int Ed Engl 36:2037
200. Fu GC, Grubbs RH (1993) J Am Chem Soc 115:3800
201. Nicolaou KC, Postema MHD, Caliborne CF (1996) J Am Chem Soc 118:1565
202. Sita LR (1995) Macromolecules 28:656
203. Coates GW, Grubbs RH (1996) J Am Chem Soc 118:229
204. Kirkland TA, Grubbs RH (1997) J Org Chem 62:7310
205. Fujimura O, Fu GC, Grubbs RH (1994) J Org Chem 59:4029
206. Fu GC, Grubbs RH (1992) J Am Chem Soc 114:5426
207. Armstrong SK, Christie BA (1996) Tet Lett 37:9373
208. Shon YS, Lee TR (1997) Tet Lett 38:1283
209. Fu GC, Grubbs RH (1992) J Am Chem Soc 114:7324
210. Martin SF, Wagman AS (1995) Tet Lett 36:1169
211. Barrett AGM, Baugh SPD, Gibson VC, Giles MR, Marshall EL, Procopiou PA (1996) Chem Commun 2231
212. Martin SF, Chen HJ, Courtney AK, Liao YS, Patzel M, Ramser MN, Wagman AS (1996) Tetrahedron 52:7251
213. Martin SF, Liao YS, Chen HJ, Patzel M, Ramser MN (1994) Tet Lett 35:6005
214. Ramza J, Descotes G, Basset JM, Mutch A (1996) J Carbohydr Chem 15:125
215. Descotes G, Ramza J, Basset JM, Pagano S (1994) Tetrahedron Lett 35:7379

216. Harrity JPA, La DS, Cefalo DR, Visser MS, Hoveyda AH (1998) J Am Chem Soc 120:2343
217. Martin SF, Liao YS, Wong YL, Rein T (1994) Tet Lett 35:691
218. Hölder S, Blechert S (1996) Synlett 505
219. Fürstner A, Langemann K (1996) J Org Chem 61:8746
220. Xu ZM, Johannes CW, Salman SS, Hoveyda AH (1996) J Am Chem Soc 118:10926
221. Houri AF, Xu ZM, Cogan DA, Hoveyda AH (1995) J Am Chem Soc 117:2943
222. Clark JS, Kettle JG (1997) Tet Lett 38:127
223. Alexander JB, La DS, Cefalo DR, Hoveyda AH, Schrock RR (1998) J Am Chem Soc 120:4041

Ruthenium-Catalyzed Metathesis Reactions in Organic Synthesis

A. Fürstner

Well-defined ruthenium carbene complexes of type 1 introduced by Grubbs et al. are among the most popular and useful metathesis catalysts known to date. They combine a high activity with an excellent tolerance towards polar functional groups and provide access to carbo- and heterocycles of almost any ring size by ring closing metathesis (RCM) of diene substrates. This includes even medium-sized and macrocyclic rings. The design and mode of action of these and related catalysts will be outlined and their performance will be illustrated by some selected applications to the synthesis of complex natural products. Moreover, some promising developments concerning a new generation of ruthenium-based catalysts (i.a. the allenylidene complexes 2), as well as recent advancements related to the use of supercritical carbon dioxide (scCO$_2$) as reaction medium for RCM, will be discussed.

Keywords: Ruthenium-carbenes, Ruthenium-allenylidenes, Ring closing metathesis, Natural product synthesis, Fine chemicals.

1
Introduction

The metathesis activity of various simple ruthenium compounds has already been discerned during early studies on the ring opening metathesis polymerization (ROMP) of norbornene and other cycloalkene substrates [1, 2]. A real breakthrough, however, was achieved when Grubbs et al. reported in the early 1990s that ruthenium-carbene complexes of the general type 1 are highly active single component (pre)catalysts for any kind of olefin metathesis reaction (Fig. 1) [3–5].

The well balanced electronic and coordinative unsaturation of their Ru(II) center accounts for the high performance and the excellent tolerance of these complexes toward an array of polar functional groups. This discovery has triggered extensive follow up work and carbenes 1 now belong to the most popular metathesis catalysts which set the standards in this field [3]. Many elegant applications to the synthesis of complex target molecules and structurally diverse natural products highlight their truely remarkable scope.

The following account describes the preparation, structure and activity of these catalysts and the insight gained into their mode of action. Based on this information, it was recently possible to design new metathesis (pre)catalysts, i.e. ruthenium allenylidene complexes of type 2, which rival the activity of 1 but are much easier to prepare [6]. This review also covers the application of these tools to advanced organic synthesis, with the focus being on the ring closing meta-

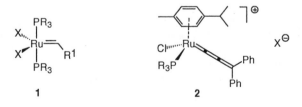

Fig. 1.

thesis (RCM) of dienes to cycloalkenes of all ring sizes. Finally, some recent investigations on metathesis reactions in supercritical carbon dioxide ($scCO_2$) will be summarized [7], since this unconventional reaction medium may eventually upgrade the industrial application profile of such transformations for the production of fine chemicals. For further applications of ruthenium catalysts (e.g. enyne metathesis, cross metathesis, polymerization reactions), however, the reader is referred to other chapters of this monograph.

2
Synthesis, Structure, Mechanism and Activity of Ruthenium-Based Metathesis Catalysts

2.1
Ruthenium Carbene Complexes and General Aspects of RCM

The preparation of the ruthenium vinylcarbenes 1a–c in the early 1990s denotes a ground breaking development in olefin metathesis in general terms [4]. These complexes contain a *late* transition metal in a *low* oxidation state [i.e. formally a neutral, 16-electron, 5-coordinate Ru(II) fragment], in contrast to most of the established metathesis catalysts at that time [1]. They are usually prepared by a metal-induced rearrangement of 3,3-diphenylcyclopropene (Scheme 1) [4, 8], but similar species can also be obtained by reaction of $Ru(H)(H_2)Cl(PCy_3)_2$ with propargyl chlorides [9]. Because 1a–c combine a high metathesis activity with a remarkable tolerance and stability toward air, water and many functional groups, they soon became attractive tools for practical applications.

In a formal sense, complexes 1 represent pre-catalysts that convert in the first turn of the catalytic cycle (vide infra) into ruthenium methylidene species of type 3 which are believed to be the actual propagating species in solution (Schemes 2, 4). The ease of formation of 3 strongly depends on the electronic properties of the original carbene moiety in 1. In addition to complexes 1a–c with $R^1=CH=CPh_2$, ruthenium carbenes with $R^1=aryl$ (e.g. 1d, Scheme 3) constitute another class of excellent metathesis pre-catalysts, which afford the methylidene complex 3 after an even shorter induction period [5]. In contrast, any kind of electron-withdrawing (e.g. -COOR) or electron-donating substitu-

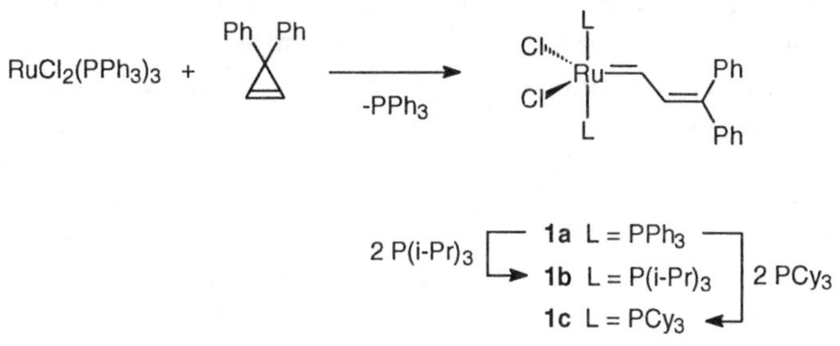

Scheme 1.

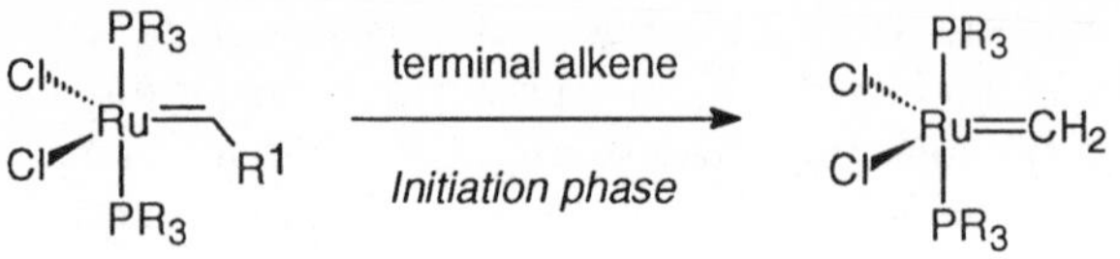

Scheme 2.

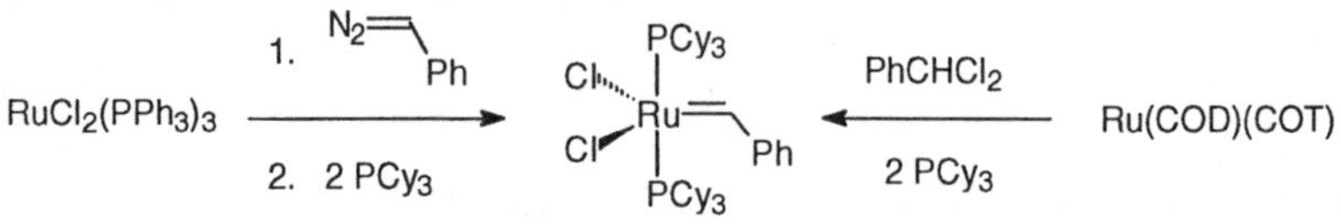

Scheme 3.

ent R^1 (e.g. -OR, -SR, -NR$_2$) impedes the formation of **3** and thus leads to a significant or even total loss of catalytic activity.

Compound **1d** can be prepared either by reaction of RuCl$_2$(PPh$_3$)$_3$ with phenyldiazomethane followed by an exchange of PPh$_3$ for PCy$_3$ [5], or by an oxidative addition of [Ru(COD)(COT)] into α,α-dichlorotoluene in the presence of PCy$_3$ (Scheme 3) [10]. These procedures avoid the somewhat laborious synthesis of 3,3-diphenylcyclopropene necessary for the preparation of the vinylcarbenes **1a–c** [4]. The fact that **1d** is now commercially available contributes greatly to the popularity of this particular metathesis "catalyst" [60].

Systematic investigations of the factors governing the catalytic activity of ruthenium carbenes of type **1** have been carried out [5, 11]. From these studies it can be deduced that electron-donating phosphines with large cone angles lead to particularly active catalysts [PPh$_3$<<P(i-Pr)$_2$Ph<PCy$_2$Ph<P(i-Pr)$_3$<PCy$_3$]. In contrast, the following order of increasing activity was determined for the anionic ligand X=I<Br<Cl, i.e. the smaller and more electron-withdrawing chloride leads to the most active species. As mentioned above, the initiation period is shorter if the substituent on the carbene moiety is R^1=C$_6$H$_4$Y rather than R^1=CH=CPh$_2$; however, electronic effects of the substituent Y on the phenyl ring turned out to be of minor importance (Y=H, NMe$_2$, OMe, Me, F, Cl, NO$_2$).

Further variations of the basic motif of **1** comprise carbene **1e** in which the anionic ligands are trifluoroacetate rather than chloride (Fig. 2) [4d]. However, this specific compound shows a significant tendency to isomerize the double bonds of the substrates in addition to its metathetic activity. Moreover, water soluble catalysts have been developed that contain bulky aliphatic phosphine ligands with either a quarternary ammonium group (**1f,g**) or a sulfonate function (**1h**) [12]. They have been tested in the ring opening polymerization

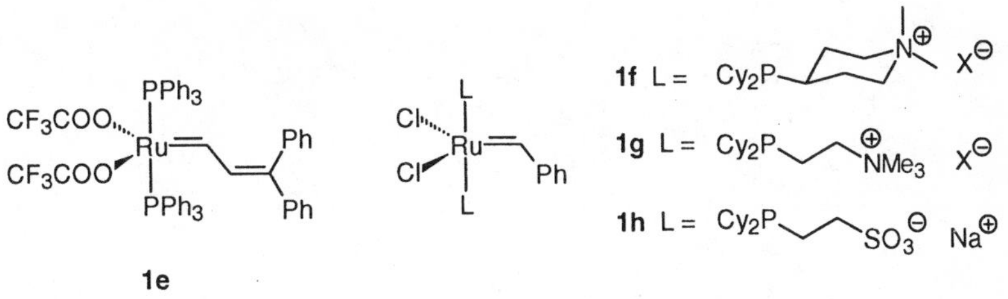

Fig. 2

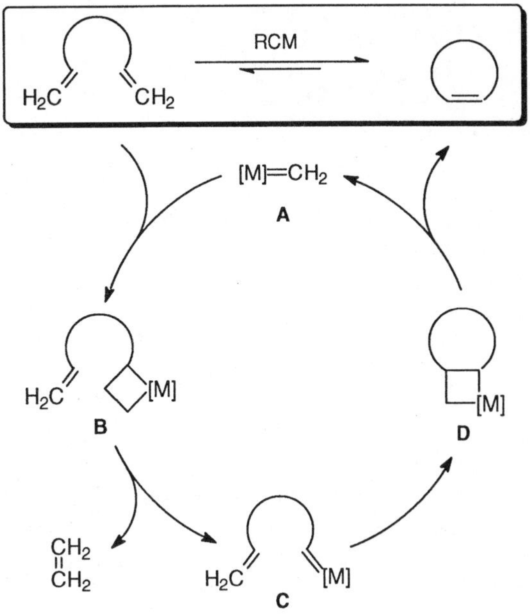

Scheme 4.

(ROMP) of functionalized oxanorbornenes in water, methanol and aqueous emulsions, but have not yet been used in RCM.

The overall mechanism of RCM is generally believed to proceed via the alternating series of formal [2+2] cycloaddition/cycloreversion steps outlined in Scheme 4 (Chauvin mechanism) [1, 13], although many details concerning the structure and properties of the intermediates A–D involved in this catalytic cycle are not yet fully resolved. Only recently, an in-depth study provided some insights into the actual nature of such intermediates [11]. This investigation suggests that a pathway involving the dissociation of one of the two phosphines bound to 1 accounts for approximately 95% of the catalyst turnover (Scheme 5). The alternative path in which both phosphine ligands remain attached to the central metal, though operative, is much less efficient.

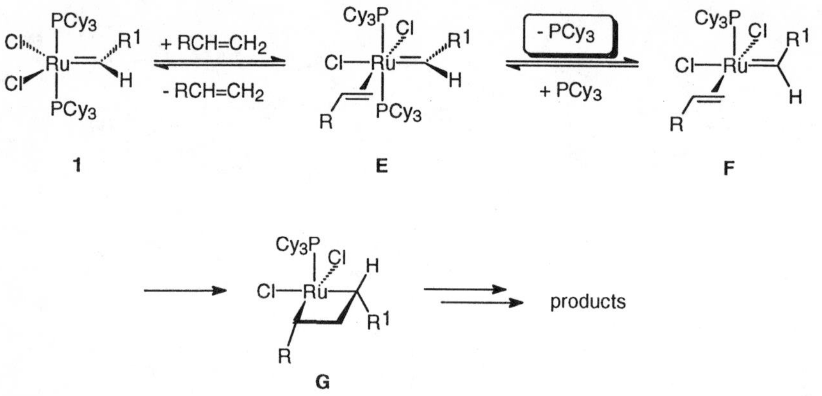

Scheme 5.

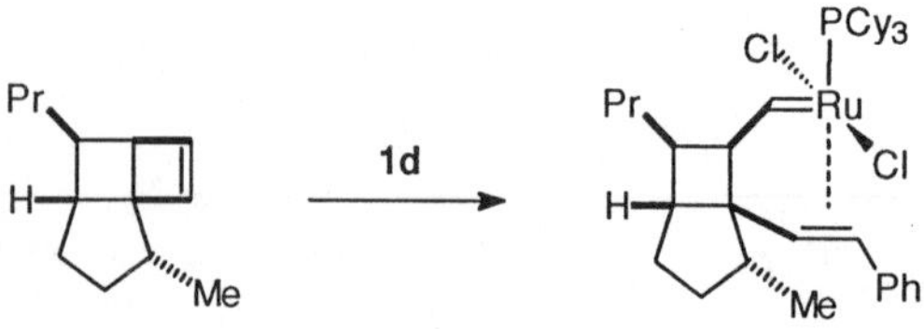

Scheme 6.

The olefin binding site is presumed to be *cis* to the carbene and *trans* to one of the chlorides. Subsequent dissociation of a phosphine paves the way for the formation of a 14-electron metallacycle **G** which upon cycloreversion generates a productive intermediate [11]. The metallacycle formation is the rate determining step. The observed reactivity pattern of the pre-catalyst outlined above and the kinetic data presently available support this mechanistic picture. The fact that the catalytic activity of ruthenium carbene complexes **1** may be significantly enhanced on addition of CuCl to the reaction mixture is also very well in line with this dissociative mechanism [11]: Cu(I) is known to trap phosphines and its presence may therefore lead to a higher concentration of the catalytically active monophosphine metal fragments **F** and **G** in solution.

Another piece of mechanistic evidence was reported by Snapper et al. [14], who describe a ruthenium catalyst "caught in action". During studies on ring opening metathesis, these authors were able to isolate and characterize carbene **5** in which a tethered alkene group has *replaced* one of the phosphines originally present in **1d**. Control experiments have shown that compound **5** by itself is catalytically active, thus making sure that it is a true intermediate of a dissociative pathway rather than a dead-end product of a metathetic process.

In all subsequent discussions it is important to note that the basic catalytic cycle of any RCM reaction (Scheme 4) is – in principle – reversible. The fact that the overall process becomes useful in preparative terms is mainly due to the following features:

1. Since RCM inevitably cuts one molecule into two, the forward reaction is entropically driven.
2. The equilibrium is constantly shifted towards the cycloalkene if ethylene or another volatile olefin is formed as the by-product.
3. If the product has a more highly substituted double bond than the substrate, the retro-reaction is kinetically hindered because most catalysts are sensitive to the substitution pattern of the olefin (cf. Sect. 3.1).
4. The bias of a given diene for RCM depends on the ring size formed, on conformational constraints of the substrate, on the presence of functional groups, and on interactions with the specific catalyst used.
5. RCM of a diene substrate can be favored over competing polymerization via acyclic diene metathesis (ADMET) by adjusting high dilution conditions.

2.2
Ruthenium Allenylidene Complexes

The ruthenium carbene complexes 1 discussed in the previous chapter have set the standards in the field of olefin metathesis and are widely appreciated tools for advanced organic synthesis [3]. A serious drawback, however, relates to the preparation of these compounds requiring either 3,3-diphenylcyclopropene or diazoalkanes, i.e. reagents which are rather difficult to make and/or fairly hazardous if used on a large scale [60]. Therefore, a need for metathesis catalysts persists that exhibit essentially the same activity and application profile as 1 but are significantly easier to make.

One class of compounds that might meet these criteria are ruthenium allenylidene complexes of the general type 2 (Fig. 1) [6]. These cationic, 18 electron ruthenium species are the first examples for the use of metal allenylidene complexes in catalysis in general. They are conveniently prepared in only two high yielding steps from commercially available and stable precursors (Scheme 7). Cleavage of well accessible dimeric precursors of the general type $[(\eta^6\text{-arene})RuCl_2]_2$ 6 on treatment with a phosphine gives access to monomeric complexes 7, which readily react with propargyl alcohols such as 8 in the presence of, e.g., $NaPF_6$ at ambient temperature to yield the desired cationic allenylidene catalysts 2 ($X=PF_6^-$). These derivatives are fairly stable and can be stored under Ar for extended periods of time without any loss of activity. Since several variations of the ligand sphere are possible, these compounds hold the promise that their activity can be fine tuned if necessary.

Based on the insight that a dissociative mechanism plays the major role along the metathesis pathway [11], these catalysts have been designed such that only *one* bulky phosphine, *one* chloride and *one* cumulenylidene ligand are attached to a Ru(II) center. Because arene ligands are known to be labile on such a metal fragment, they will easily liberate free coordination sites (❑) for the interaction with the alkene substrate. Although the precise mode of action of such allenyli-

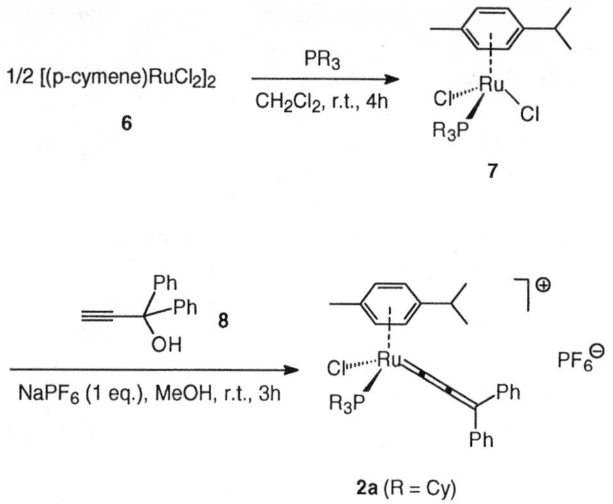

Scheme 7.

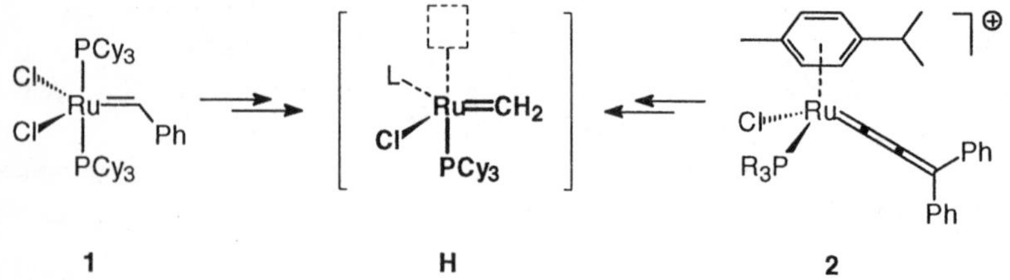

Scheme 8.

dene complexes remains to be elucidated, it is likely that they lead in solution to a catalytically active metal fragment **H** which is similar to the one formed from Grubbs-type carbenes **1**; the essential difference resides in the nature of the second ionic ligand L (Scheme 8). The fact that **1** and **2** show a very similar correlation of their catalytic activity with the nature of the phosphine ligand [PCy_3>$P(i-Pr)_3$>>PPh_3] corroborates this assumption [6].

The direct comparison of **1** and **2** in a variety of RCM reactions also indicates a presumably close relationship between these catalysts (Table 1) [6]. Both of them give ready access to cycloalkenes of almost any ring size ≥ 5, including medium sized and macrocyclic products. Only in the case of the 10-membered jasmine ketolactone **16** was the yield obtained with **2a** lower than that with **1c**; this result may be due to a somewhat shorter lifetime of the cationic species in solution. However, the examples summarized in Table 1 demonstrate that the allenylidene species **2** exhibit a remarkable compatibility with polar functional groups in the substrates, including ethers, esters, amides, sulfonamides, ketones, acetals, glycosides and even free hydroxyl groups.

Table 1. RCM employing Ru-allenylidene catalyst **2a**[a]; comparison with literature data using Ru-carbene **1c** as the catalyst [6]

Substrate	Product	Yield[b]	
		2a	**1**
9	**10**	83%[c]	93%
11	**12**	86%	90% 62%
13	**14**	75%	68%
15	**16**	40%	86%
17	**18**	90%	79%
19	**20**	73%	83%
21	**22**	85%	77%

[a] All reactions using **2a** were carried out in toluene at 80 °C using a catalyst loading of 5 mol% unless stated otherwise.
[b] Isolated yield.
[c] With only 2.5 mol% of **2a**.

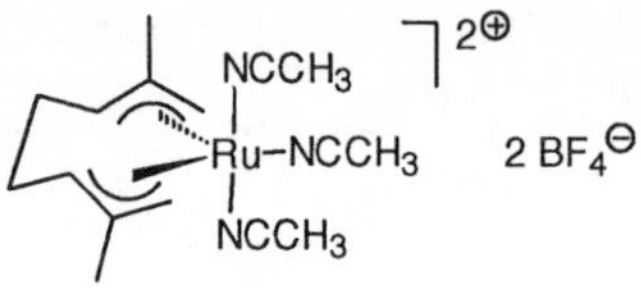

23

Fig. 3

2.3
Less Defined Ruthenium Catalysts

It has been shown that $[(\eta^6\text{-arene})RuCl_2]_2$ **6** and $[(\eta^6\text{-arene})RuCl_2]\cdot PR_3$ **7** complexes can be activated in situ to afford active metathesis catalysts, either on treatment with diazoalkanes [15] or by UV irradiation [16]. The structure of the active species thus formed is unknown, but it initiates the ring opening metathesis polymerization reactions (ROMP) of various cycloalkenes very efficiently. Therefore these in situ recipes may also be useful in the context of preparative organic chemistry.

Allylruthenium(IV) complexes such as **23** convert into highly performing metathesis catalysts on treatment with ethyl diazoacetate (Fig. 3) [17]. Again, the structure of the active species is unknown and only applications to polymerization reactions have been reported so far.

3
The Application Profile of the Standard Ruthenium-Based Metathesis Catalysts in Synthesis

Since the vinylcarbenes **1a–c** and the aryl substituted carbene (pre)catalyst **1d**, in the first turn of the catalytic cycle, both afford methylidene complex **3** as the propagating species in solution, their application profiles are essentially identical. Differences in the *rate* of initiation are relevant in polymerization reactions, but are of minor importance for RCM to which this chapter is confined. Moreover, the close relationship between **1** and the ruthenium allenylidene complexes **2** mentioned above suggests that the scope and limitations of these latter catalysts will also be quite similar. Although this aspect merits further investigations, the data compiled in Table 1 clearly support this view.

3.1
Effects of Olefin Substitution

The ruthenium compounds described above show a distinctly lower metathetic activity than the molybdenum alkenylidene complex **24** developed by Schrock et al. (Fig. 4, see also the chapter by R.R. Schrock, this volume) [18], which is another standard catalyst for any type of olefin metathesis reaction. However, they

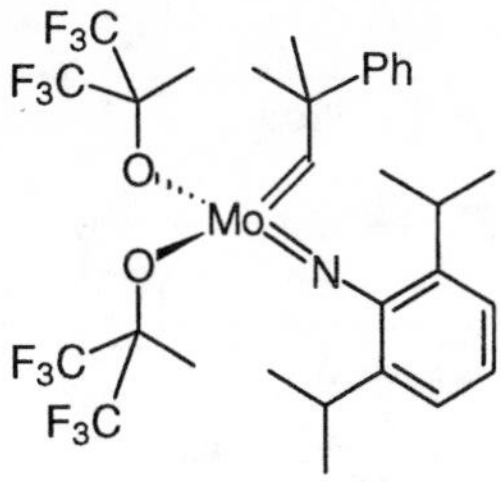

24

Fig. 4

compensate for this lower intrinsic reactivity by an increased tolerance towards functional groups and a somewhat higher selectivity (cf. Sect. 3.2).

In particular, ruthenium carbenes **1** are more sensitive to the substitution pattern of the alkenes than the molybdenum catalyst **24** [19]. While the latter reacts readily even with di- and tri-substituted double bonds and is apparently the only catalyst capable of producing tetrasubstituted cycloalkenes (cf. Table 2, en-

Table 2. Formation of tri- and tetrasubstituted alkenes by RCM; comparison of the efficiency of catalysts **1** and **24** (E=COOMe)[a] [19]

No.	Substrate	Product	Yield	
			1d	**24**
1			93%	[100%]
2	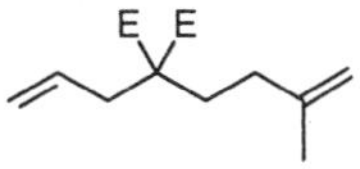	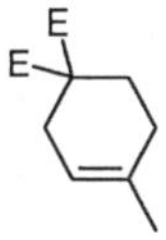	97%	[100%]
3	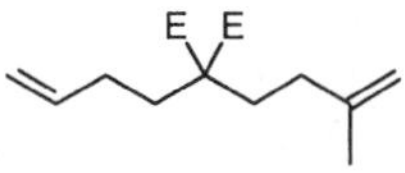	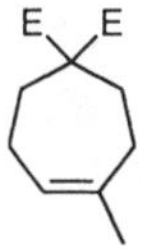	96%	[100%]
4	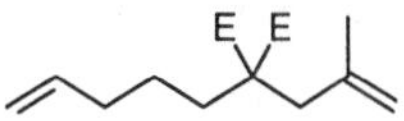	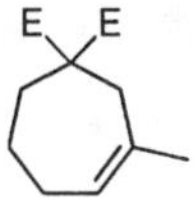	[25%]	96%
5	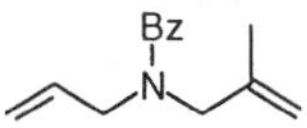		97%	[100%]

[a] Isolated yield [yield determined by NMR].

Table 2. continued

No.	Substrate	Product	Yield	
			1d	**24**
6			0%	93%
7			0%	61%
8			98%	[100%]
9			0%	96%
10			[25%]	97%
11			[5%]	89%
12			0%	0%
13			98%	decomp.
14			97%	[100%]

[a] Isolated yield [yield determined by NMR].

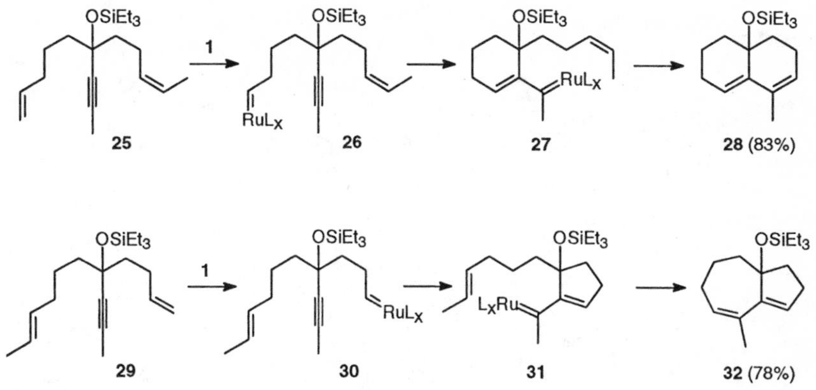

Scheme 9.

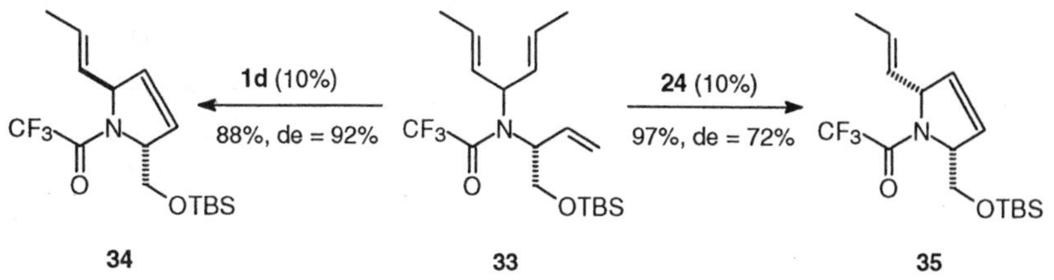

Scheme 10.

tries 6, 7), the best substrates for reactions with **1** and **2** are terminal, monosubstituted olefins. However, **1** may eventually lead to the formation of trisubstituted olefins in kinetically favorable cases when 5-, 6- or 7-membered rings are formed as shown by the examples compiled in Table 2.

In turn, the propensity of **1** to respond to steric hindrance can be used to control the site of initiation of an RCM reaction in a polyene substrate (Scheme 9) [20]. Thus, dienyne **25** reacts with the catalyst regioselectively at the least substituted site; the evolving ruthenium carbene **26** undergoes a subsequent enyne metathesis leading to a new carbene **27**, which is finally trapped by the disubstituted olefin to afford the bicyclo[4.4.0]decadiene product **28**. By simply reversing the substitution pattern of the double bonds, the complementary bicyclo [5.3.0] compound **32** is formed exclusively, because the cyclization cascade is then triggered at the other end of the substrate. Note that in both examples trisubstituted olefins are obtained by means of a ruthenium based metathesis catalyst [20]!

The idea of determining the site of initiation via the substitution pattern of the olefin has also been used by Blechert et al. during the course of a stereocontrolled RCM process (Scheme 10). Again, the reaction starts most likely at the terminal olefin site in **33** independent of whether **1** or **24** is used as catalyst; however, due to the different coordination geometries of ruthenium and molybdenum, the evolving carbene reacts with either diastereotopic olefin attached to

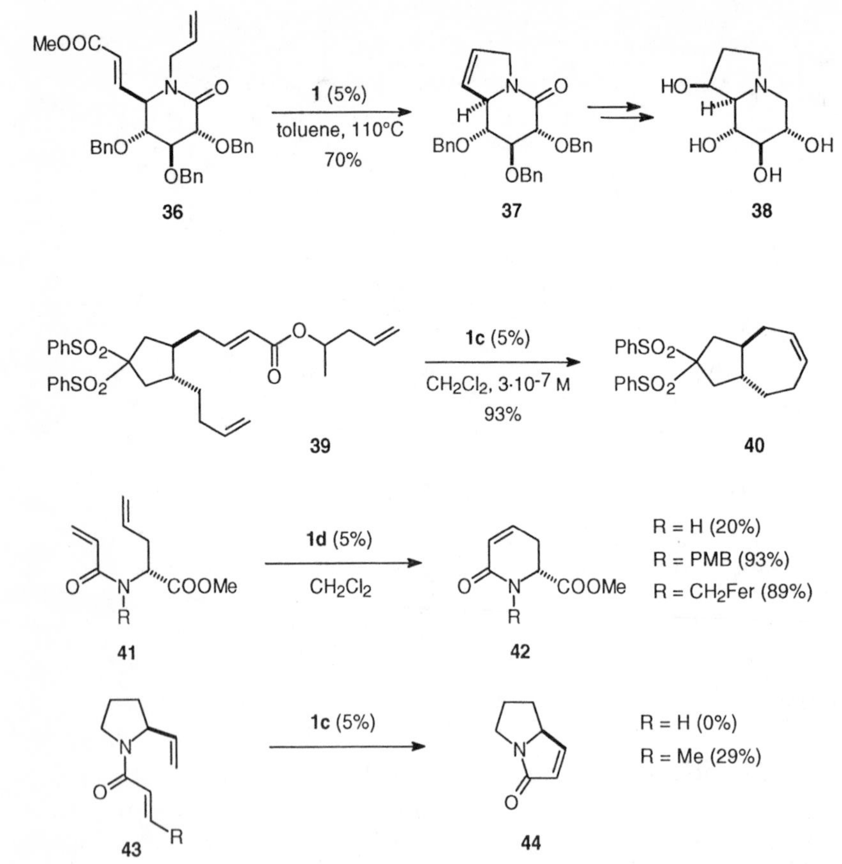

Scheme 11.

the nitrogen substituent, thus leading to the selective formation of **34** or **35**, respectively (Scheme 10) [21].

Polar substituents on the double bonds of a given substrate – both electron donating and electron withdrawing ones – usually stop the catalytic process [5d]. Among the (few) exceptions to this rule are cyclization reactions involving α,β-unsaturated esters or amides (Scheme 11) as exemplified by the formal total synthesis of the glucosidase inhibitor castanospermine **38** [22] and by the formation of compound **40** starting from triene **39**. The latter transformation is the only case reported so far in which an α,β-unsaturated ester reacts preferably over a terminal alkene [23]. This is likely due to the proximity of the reacting sites and to the kinetically favored formation of a 7- rather than a 13-membered ring. In addition, various successful reactions involving α,β-unsaturated amides have been reported in which the carbonyl group is not removed during ring closure (e.g. **41**→**42**) [24]. However, the application of this strategy to the synthesis of (S)-pyrrolactam A **44** gave only rather poor yields [25].

3.2
Functional Group Compatibility

The ruthenium carbene catalysts 1 developed by Grubbs are distinguished by an exceptional tolerance towards polar functional groups [3]. Although generalizations are difficult and further experimental data are necessary in order to obtain a fully comprehensive picture, some trends may be deduced from the literature reports. Thus, many examples indicate that ethers, silyl ethers, acetals, esters, amides, carbamates, sulfonamides, silanes and various heterocyclic entities do not disturb. Moreover, ketones and even aldehyde functions are compatible, in contrast to reactions catalyzed by the molybdenum alkylidene complex 24 which is known to react with these groups under certain conditions [26]. Even unprotected alcohols and free carboxylic acids seem to be tolerated by 1. It should also be emphasized that the sensitivity of 1 toward the substitution pattern of alkenes outlined above usually leaves pre-existing di-, tri- and tetrasubstituted double bonds in the substrates unaffected. A nice example that illustrates many of these features is the clean dimerization of FK-506 45 to compound 46 reported by Schreiber et al. (Scheme 12) [27].

Problematic functional groups, however, are thioethers and disulfides [28] as well as free amines which poison catalysts of type 1 [4c]. In case of amines this problem is easily solved by choosing either an appropriate protecting group for nitrogen (e.g. amide, sulfonamide, urethane), or simply by protonation since ammonium salts were found to be compatible with 1 [4c]. As will be discussed in Sect. 4, free amines can also be metathesized in supercritical CO_2 as the reaction medium [7].

Although the ruthenium allenylidene complexes 2 have not yet been as comprehensively studied as their carbene counterparts, they also seem to exhibit a closely related application profile [6]. So far, they have proven to tolerate ethers, esters, amides, sulfonamides, ketones, acetals, glycosides and free secondary hydroxyl groups in the substrates (Table 1).

This excellent compatibility of ruthenium catalysts encourages the use of RCM as a key strategic element for the synthesis of complex target molecules as will become evident from the examples discussed in Sect. 5.

3.3
Ring Size

Successful applications of RCM to the formation of almost any ring size ≥ 5 have been reported in the literature, including medium-sized and macrocyclic compounds. While the synthesis of 5-, 6- and 7-membered rings is rather general [3] (for representative examples see Table 3) and only few failures have been recorded, unfavorable conformational effects and/or strain in the molecule may render the cyclization of larger rings significantly more delicate. Therefore it is appropriate to discuss the present state of the art in this particular field in some detail.

The formation of medium sized rings is a formidable challenge, since it is well established that the inverse reaction, i.e. the ring opening metathesis polymeri-

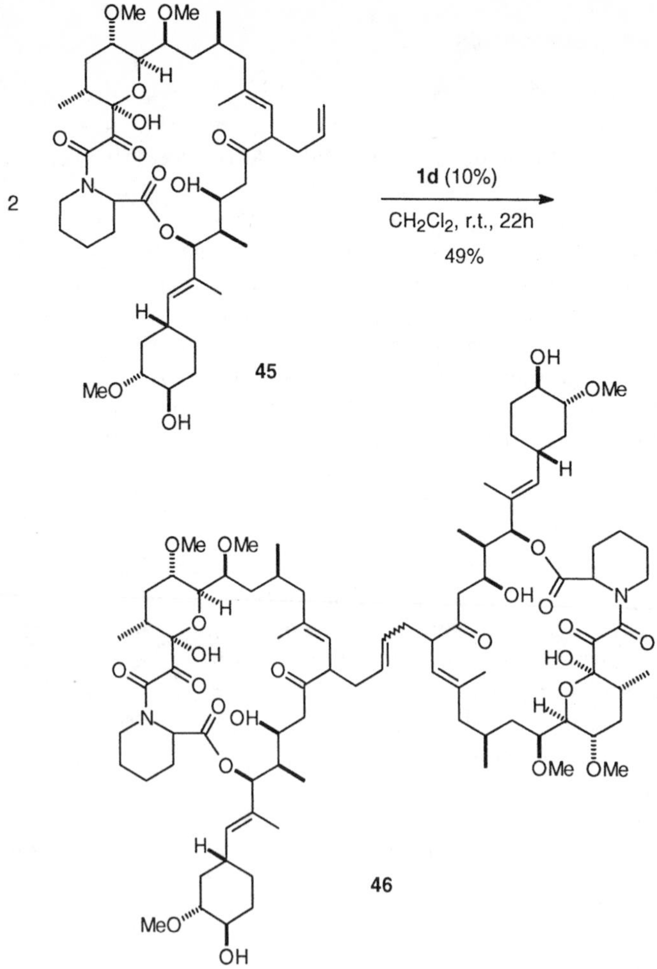

Scheme 12.

zation (ROMP) of cycloalkenes with 8–11 ring atoms, is a highly favored process due to the release of ring strain [1]. Despite this seemingly adverse situation, a rapidly growing number of successful applications indicates that even medium sized rings can be prepared by RCM in moderate to excellent yields (Tables 1, 4) [29]. It is likely that the entropy gained upon dissecting *one* diene into *two* olefins together with the evaporative loss of ethylene (cf. Scheme 4) can compensate for the unfavorable increase in enthalpy during the cyclization process. However, subtle changes in the substrates can have a major impact on the outcome of the reaction. This is evident from a comparison of the different 8-membered rings 47–54 compiled in Scheme 13 [29i, j]. It is generally accepted that some conformational predisposition towards ring closure in the diene substrates – as induced, e.g. by annelation, the Thorpe-Ingold effect, hydrogen bonding or re-

Table 3. Representative examples of 5-, 6- and 7-membered rings formed by RCM using 1 as catalyst

Product (Yield)	Ref.	Product (Yield)	Ref.
0% (n = 1) 89% (n = 2) 59% (n = 3)	[49]	52%	[50]
Ts—N 83% P = polymer	[24b]	94%	[51]
97%	[46]	—OBn 40%	[52]
95%	[53]	98%	[54]
89%	[4c]	85-90%	[55]
TBSO 90%	[56]	82%	[57]
93%	[24a]	78%	[58]

lated constraints – is required to make such reactions effective although we are still far away from a clear picture.

In the case of macrocyclic rings, the situation is better understood. In contrast to earlier statements in the literature [3a], even diene substrates devoid of any conformational pre-disposition towards ring closure turned out to be excellent substrates for macrocylization reactions catalyzed by ruthenium-carbene or -allenylidene complexes. From these investigations [30], however, a set of parameters has been deduced which turned out to be decisive:

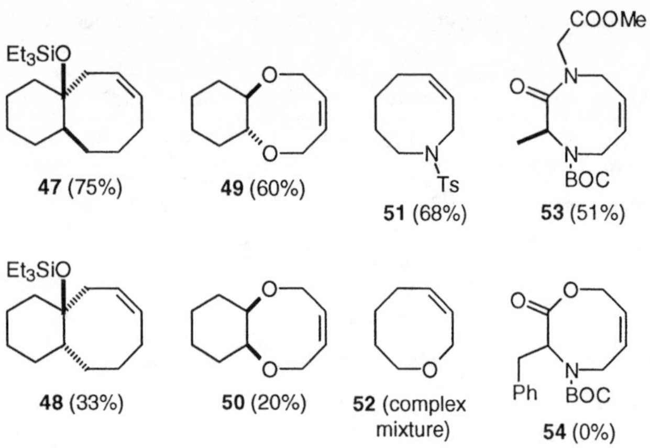

Et3SiO
47 (75%)
49 (60%)
51 (68%)
COOMe
BOC
53 (51%)
Et3SiO
48 (33%)
50 (20%)
52 (complex mixture)
Ph
BOC
54 (0%)

Scheme 13.

Table 4. Selected examples of medium-sized rings formed by RCM with **1** as catalyst

Product (Yield)	Ref.	Product (Yield)	Ref.
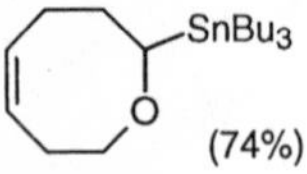 (74%)	[29j]	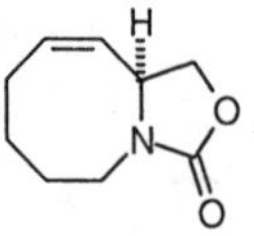 (75%)	[29e]
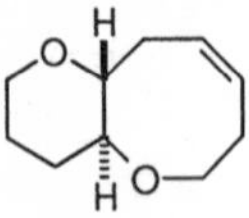 (>95%)	[59]	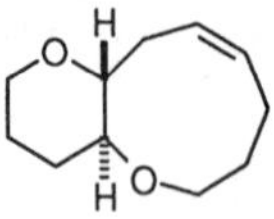 (45%)	[59]
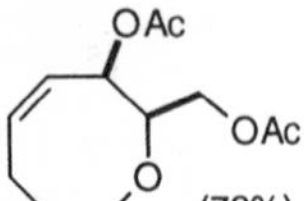 (73%)	[29k]	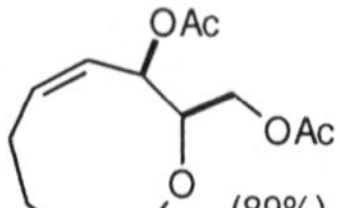 (89%)	[29k]
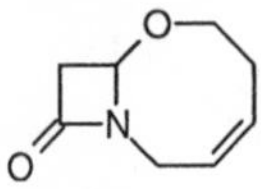 (53%)	[29n]	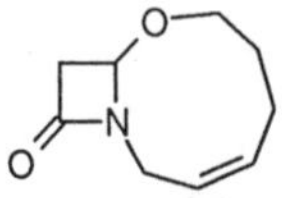 (12%)	[29n]
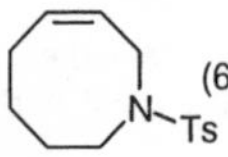 (68%)	[29f]	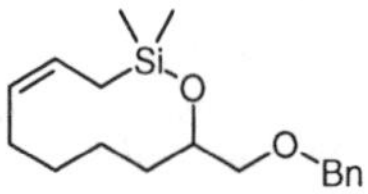 (73%)	[29m]
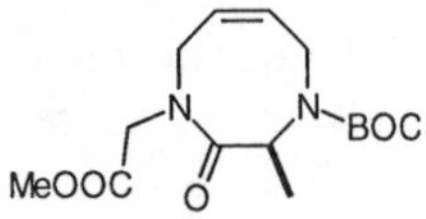 (51%)	[29h]	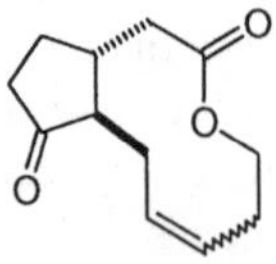 (88%)	[29a]

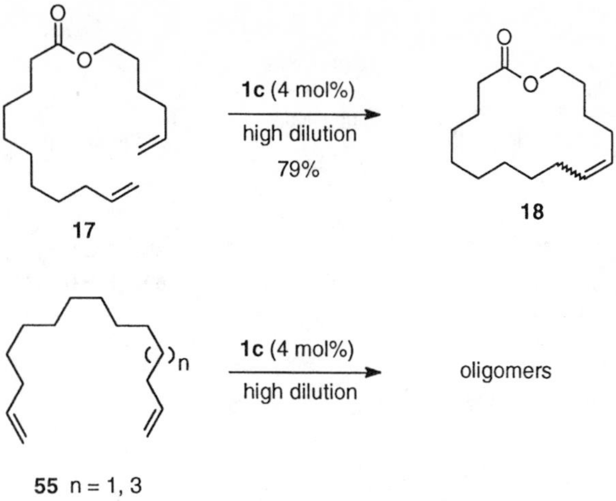

Scheme 14.

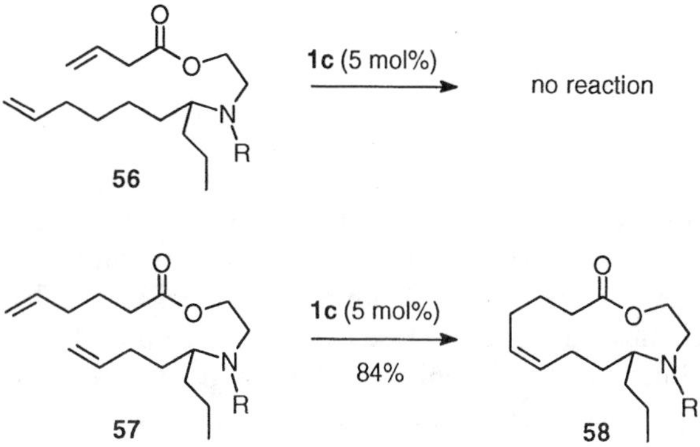

Scheme 15.

1. The mere presence of a polar functional group (ester, amide, ketone, ether, sulfonamide, urethane, etc.) seems to be a fundamental requirement for smooth macrocyclizations by RCM. The two examples displayed in Scheme 14 nicely illustrate this aspect.
2. The site of ring closure is another key issue as indicated by Scheme 15.
3. Steric hindrance close to the double bonds significantly lowers the yield (cf. **60a/b,** Scheme 16).

These observations may be rationalized by assuming that the polar functional group coordinates to the metal center in one or more intermediates along the RCM pathway (Scheme 17) [30b]. Such a Lewis-acid/Lewis-base interaction may assemble the reacting sites within the coordination sphere of the ruthenium and hence provide internal bias for cyclization (e.g. structure I). However, if such an

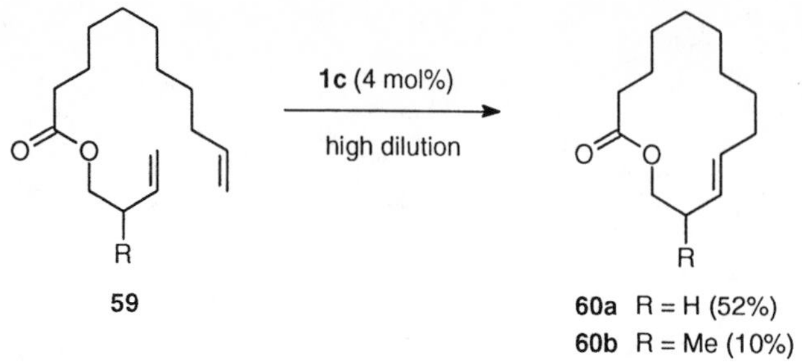

Scheme 16.

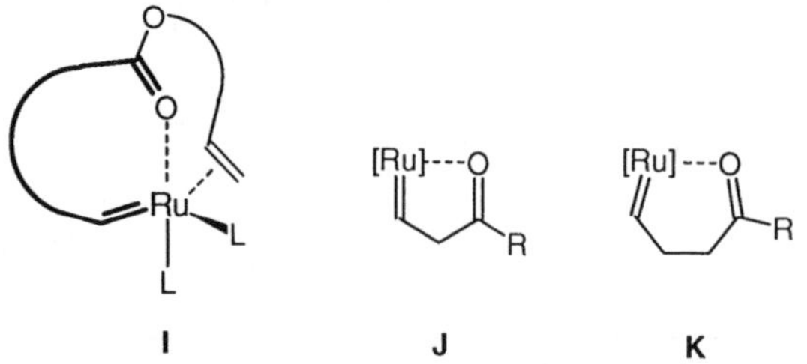

Scheme 17.

intermediate becomes too stable, as might be the case with 5- or 6-membered chelate structures such as **J** or **K**, the catalyst is sequestered in an unproductive form and the conversion ceases. Taking into account that the residual ligands L in **I** (at least one PCy_3) are very space filling, this model also explains the adverse effects of bulky substituents at or close to the double bonds to be metathesized. Although this interpretation is only based on indirect evidence and needs further support from mechanistic investigations, this "relay model" provides a safe guidance for retrosynthetic planning.

Importantly, however, it should be noted that RCM generally leads to a mixture of the (E)- and (Z)-isomers if cycloalkenes with ≥ 10 ring atoms are formed. At present we are neither able to predict nor to control the configuration of the newly formed double bond and seemingly small variations in the substrates can have a major impact on the diasteromeric ratio. Many examples in the epothilone- [35] (see also chapter by K.C. Nicolaou et al., this volume) and azamacrolide series [30b] highlight this aspect. More details on the actual structure of the intermediates of an RCM reaction must be gathered before a general solution to this inherently difficult but preparatively most important problem can be envisaged.

The examples summarized in Table 5 and in Sect. 5 of this review illustrate the applicability of RCM to the preparation of various macrocyclic perfume ingredients [30], pheromones [30], antibiotics [31–35], crown ethers [36], cyclic peptides [37], catenanes [38] and capped calixarenes [39].

In order to properly assess the scope of such RCM-based macrocyclizations, however, it is important to comment further on the effect of conformational con-

Table 5. Selected examples of macrocycles formed by RCM using **1** as the catalyst

Product (Yield)	Reference	Product (Yield)	Ref.
(84%)	[30b]	(80%)	[30b]
(71%)	[30]	(70%)	[36b]
(80%)	[37]	(72%)	[31]
(57%)	[39]		[38]

straints. The deductions summarized above do not mean that any macrocyclic system can be formed by RCM. Although they clearly show that conformational predisposition towards ring closure is not a basic requirement for a successful macrocyclization [30], one has to keep in mind that the ability of RCM to build up strain in a molecule is limited. In other words, if a given diene is held in a highly unfavorable conformation and RCM must override a considerable increase in enthalpy during ring closure, the reaction is likely to fail.

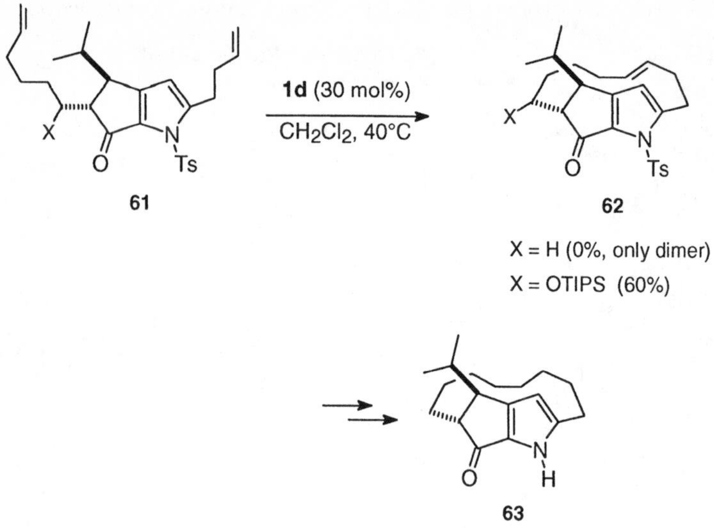

Scheme 18.

A good example for such a situation is a recent report on the synthesis of the macrotricyclic core **63** of roseophilin [40, 41]. RCM was able to form the rather strained ansa chain of this target molecule only after the cyclization had been biased by a conformational control element X which helps to bring the unsaturated chains closer together and lowers the enthalpic barrier during ring formation (Scheme 18).

3.4
Sequential Transformations

Since any metathetic conversion of a (di)olefin leads to the formation of a new olefin, this reaction can be easily incorporated into sequential transformations. An early example is the combination of an enthalpically driven ring-opening reaction of a strained cycloalkene with an entropically driven RCM step as reported by Grubbs et al. (Scheme 19) [20b]. In this specific example, the domino process is likely triggered at one of the allylic side chains of **64**. The resulting carbene **65** opens the tethered cyclobutene unit, transmits the reacting site and leads to the formation of a second dihydrofuran ring upon reaction of **66** with the remaining unsaturated appendage.

Several other sequential transformations incorporating metathesis reactions have been reported by Hoveyda (for a comprehensive review see chapter by A.H. Hoveyda in this monograph) [42] and others [43, 44].

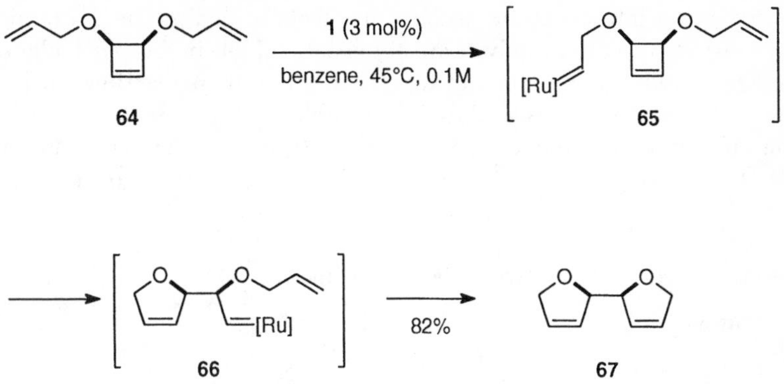

Scheme 19.

3.5
Metathesis Reactions in Solid Phase Synthesis

Medicinal chemistry makes use of solid phase combinatorial chemistry as a (rapidly maturing) tool for lead discovery and optimization. Metathesis reactions are obviously useful in this context [45] as they can be used:

1. To bind alkene substrates to a resin
2. To cleave the products of solid phase syntheses off the resin
3. To cyclize resin bound diene substrates by RCM; in this specific case, the "pseudo" dilution conditions within the polystyrene matrix may favor ring closure over competing polymerization of the substrate via ADMET
4. To effect "traceless" reactions on a support, as, e.g., exemplified by the ring closing metathesis reactions of bound dienes which concomitantly release the products

Selected examples of these and related applications are authoritatively discussed in the chapters by K.C. Nicolaou et al. (epothilone libraries) and S.E. Gibson and S.P. Keen (cross metathesis processes on solid phase) in this monograph. For some further advancements, the reader is referred to the recent literature [45].

4
Olefin Metathesis in Supercritical Carbon Dioxide

Although numerous advantages are associated with the use of supercritical carbon dioxide (scCO$_2$) as an ecologically benign and user friendly reaction medium, systematic applications to metal-catalyzed processes are still rare. A notable exception is a recent report on the use of scCO$_2$ for the formation of industrially relevant polymers by ROMP and the cyclization of various dienes or enynes via RCM [7]. Both Schrock's molybdenum alkylidene complex 24 and the ruthenium carbenes **1** introduced by Grubbs turned out to catalyze such reactions in scCO$_2$; however, the latter catalysts are not visibly soluble in this particular reaction medium and it is therefore not yet clear whether these reactions are homogeneously or heterogeneously catalyzed processes.

The efficiency and application profile of RCM in scCO$_2$ are noteworthy (Table 6). Deserving particular mention is the ease of work-up of the reaction

Table 6. RCM in scCO$_2$ catalyzed by **1** unless stated otherwise[a] [7]

No.	Substrate	Product	Yield
1			93%
2			62%
3			62% [b]
4			88%
5			67%
6			74%

[a] Catalyst **1** (1 mol%), T=313 K, d≥0.76 g cm^{-3}.
[b] Using **24** (5 mol%) as the catalyst.

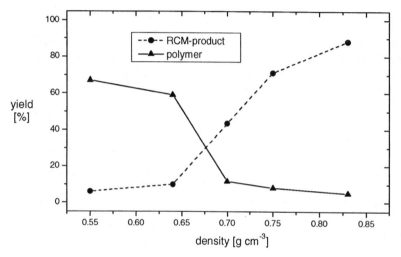

Scheme 20.

mixtures upon venting the reactor. Thus, volatile compounds such as musk-odored macrolides (Table 6, entries 4, 5) are selectively removed from the auto-clave by virtue of the extractive properties of CO_2 and can be collected in cold traps. The metal residue remaining in the reactor was found to be still active and can be used again; of course, the CO_2 can also be recycled, if desirable. These features may well upgrade the industrial application profile of RCM for fine chemical production [7].

An observation of rather fundamental importance is the remarkable and unprecedented influence of the density of the reaction medium on the path of a metathetic conversion (Scheme 20). Thus, it was found that diene **17** is selectively cyclized to the 16-membered ring **18** at densities of the compressed phase of $d \geq 0.65$ g cm^{-3}, whereas mainly oligomers are formed below this threshold [7]. Although the reasons for this effect may be significantly more complex and are far from being fully understood, Scheme 21 provides a mnemonic explanation: an increasing density of the $scCO_2$ results in a higher population of inert "sol-

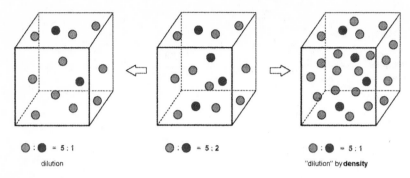

Scheme 21.

vent" molecules per volume. This effect ("dilution by density") mimics the dilution of a given substrate in a conventional solvent and favors the unimolecular process over the competing intermolecular reaction simply by making the reactive encounters of two substrate molecules less likely.

Finally, it should be mentioned that $scCO_2$ can be used as a "protective" medium for certain functional groups. This was noticed during the formation of epilachnene (Table 6, entry 6), an azamacrolide alkaloid found in the defense secretions of the pupae of the Mexican beetle, *Epilachnar varivestis*. In conventional solvents this reaction fails as the catalyst is poisoned by the amine and can only be achieved after suitable protection of the basic nitrogen (e.g. Boc, Fmoc) [30b]. However, in $scCO_2$ as the reaction medium, even the free amine cyclizes readily to the desired macrocycle (Table 6, entry 6), most likely because of the reversible formation of the corresponding carbamic acid under the reaction conditions [7].

5
Selected Applications of RCM to Target Oriented Synthesis

A proliferating number of applications of RCM to the synthesis of complex target molecules and natural products substantiates its relevance for preparative organic chemistry and illustrates the remarkable tolerance of the standard metathesis catalysts toward an array of sensitive functional groups [3]. However, RCM can even be used in a more strategic way: Since it allows to selectively activate an olefin under notably mild conditions in the presence of various other functional groups, the number of "unproductive" protection/deprotection steps and functional group interconversions can be reduced to a minimum. If properly assessed, this concept may significantly upgrade the efficiency, flexibility, practicality, "atom economy" and "economy of steps" of multiplex syntheses.

The following treatise is selective rather than comprehensive and intends to highlight some of the advantages mentioned above. The reader is also referred to previous sections of this review as well as to other chapters in this monograph which compile many additional examples.

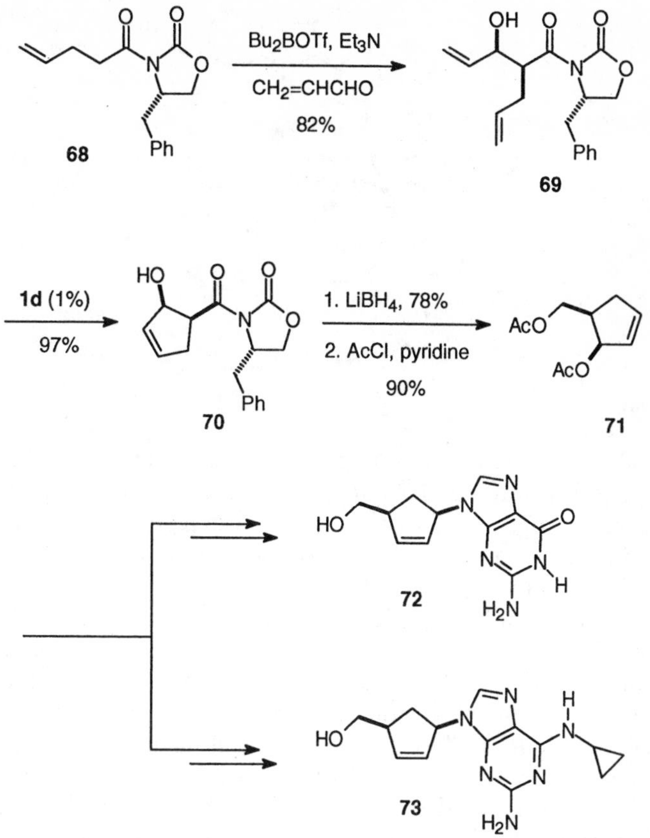

Scheme 22.

5.1
Carbocyclic Nucleosides

Carbocyclic analogues of nucleosides have attracted much attention as potential antiviral and antitumor therapeutic agents. Carbovir **72** is one of the most famous derivatives of this series, but its closely related analogue 1592U89 **73** also holds remarkable promise for the treatment of AIDS and is currently in phase II clinical trials.

A nice and convergent approach to both compounds makes use of RCM to form the 5-membered building block **71**, which mimics the carbohydrate part of the nucleosides. The necessary diene precursor **69** is readily assembled via Evans aldol chemistry. RCM then affords the ring in almost quantitative yield (**69**→**70**), leaving the chiral centers and the free hydroxyl group intact. Removal of the chiral auxiliary by reductive cleavage, attachment of the base by means of π-allylpalladium chemistry, and a final deprotection step complete these highly efficient syntheses [46].

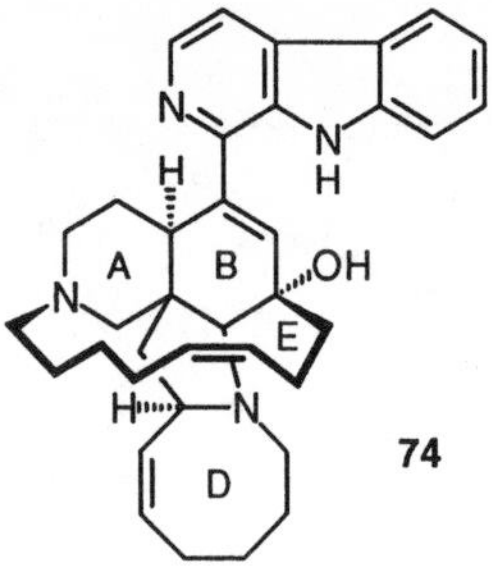

Scheme 23.

5.2
Manzamine A

The polycyclic structure of manzamine A **74**, an alkaloid with promising anti-tumor activity, constitutes an ideal testing ground for probing the effciency of RCM. Although no total synthesis of **74** has yet been reported, various approaches to this complex target rely on RCM-based strategies.

Specifically, Pandit et al. reported the formation of the 13-membered E-ring in a fairly advanced model study [34a]. Although this particular cyclization reaction gave only 30% yield after a prolonged reaction time, it was one of the first successful applications of RCM to the synthesis of macrocycles starting from diene substrates that are conformationally predisposed to ring closure by a rigid backbone. Other model studies indicate that the 8-membered D-ring of **74** can also be accessed by RCM using either **1** [29e] or Schrock's molybdenum alkylidene complex **24** as the catalyst [29d]. Finally, a recent study directed towards the ABE tricyclic core of manzamine shows a nice way to use a quarternary – and hence non-basic – pyridinium salt as an A-ring mimic during the cyclization of the macrocyclic E-ring [34b]. This circumvents the inherent problem posed by manzamine that the basic nitrogen functions in the substrate may poison the metathesis catalysts.

5.3
Epothilone A

The synthesis of the promising anti-cancer agent epothilone A **75** and many analogues thereof by means of RCM is comprehensively reviewed in the chapter by K.C. Nicolaou et al., this volume, and will not be duplicated here [35].

The only aspect to which the reader is referred is the fact that a comparison of the different strategies nicely illustrates the fundamental requirements for productive RCM as discussed in Sect. 3.3. Specifically, it is of utmost importance that (1) the distance of the olefinic sites to the polar "relay" substituents, and (2) the low steric hindrance close to the double bonds are properly assessed when choosing the site of ring closure (Scheme 24). Moreover, in all syntheses directed towards **75** or analogues thereof, mixtures of the (E)- and (Z)-isomers have been

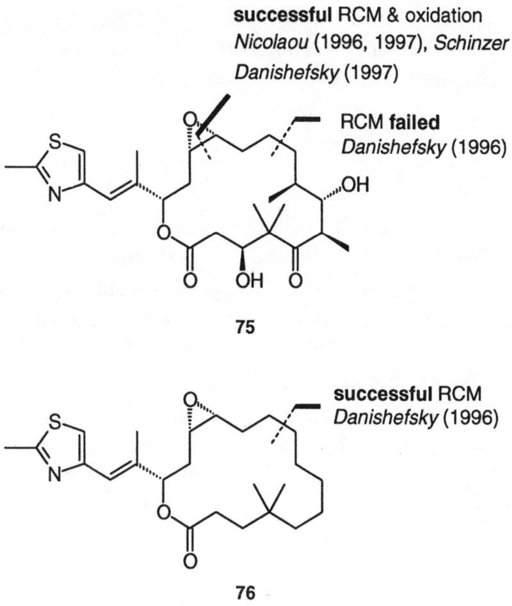
successful RCM & oxidation
Nicolaou (1996, 1997), Schinzer (1997), Danishefsky (1997)
RCM failed
Danishefsky (1996)
S
N
O
OH
O
OH
O
75
S
N
O
O
successful RCM
Danishefsky (1996)
76

Scheme 24.

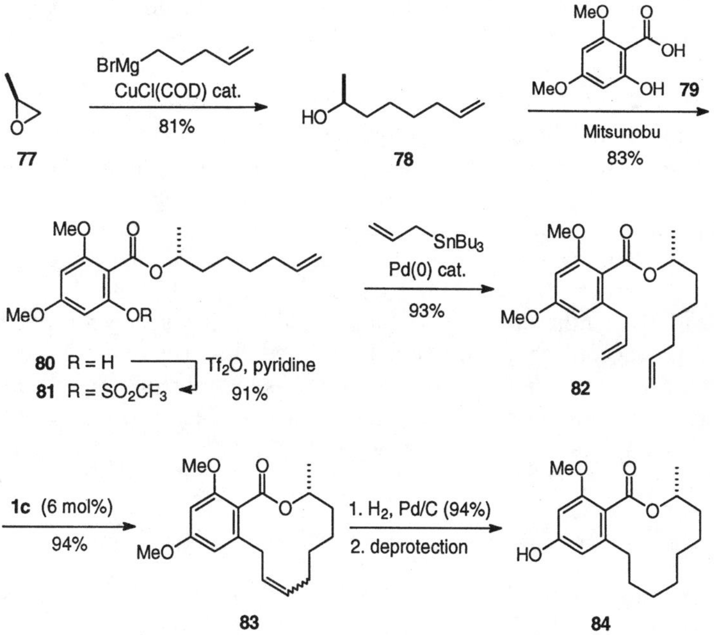
BrMg
CuCl(COD) cat.
81%
O
77
HO
78
MeO
O
OH
MeO
OH
79
Mitsunobu
83%
MeO
O
O
MeO
OR
80 R = H
81 R = SO2CF3
Tf2O, pyridine
91%
SnBu3
Pd(0) cat.
93%
MeO
O
O
MeO
82
1c (6 mol%)
94%
MeO
O
O
MeO
83
1. H2, Pd/C (94%)
2. deprotection
MeO
O
O
HO
84

Scheme 25.

formed [35], thus shedding light on the notion that we are presently neither able to control nor even to predict the stereochemistry of the newly formed double bond of large ring cycloalkenes generated by RCM [61].

5.4
(+)-Lasiodiplodin

The orsellinic acid derivative lasiodiplodin **84** and its de-*O*-methyl congener are found, i.a., in the roots of *Arnebia euchroma*, a plant which is used in traditional Chinese medicine. These compounds elicit diverse biological responses as inhibitors of prostaglandin biosynthesis, cyctotoxic agents, and plant growth regulators.

Enantiomerically pure **84** has been obtained in only seven synthetic operations in 40% *overall* yield starting from simple and commercially available precursors (Scheme 25). All C-C bond formations are metal-assisted or metal-catalyzed, with the RCM-based cyclization of the 12-membered ring (**82→83**) being particularly efficient [33].

5.5
Tricolorin A

The synthesis of the disaccharide subunit **85** of tricolorin A, a cytotoxic resin glycoside isolated from *Ipomoea tricolor*, provides a unique opportunity to compare the efficiency of an RCM-based macrocyclization reaction with that of a more conventional macrolactonization strategy. Furthermore, this specific target molecule challenges the compatibility of the catalysts with various functional groups.

Both aspects are very well met by the Grubbs carbene catalyst **1c** [32] as well as by the ruthenium allenylidene complexes **2a** [6], which effect the formation of the 19-membered lactone ring spanning two monosaccharide units in excellent yields (Scheme 26). Again, a mixture of (E)- and (Z)-isomers was formed which was hydrogenated to afford the desired building block **85**. The direct comparison of this metathesis route [6,32] with the macrolactonization protocols reported in the literature [47] confirms the notion that RCM rivals all established methods for the synthesis of large rings, provided that the proper site of ring closure is chosen according to the rationale outlined in Sect. 3.3. Of particular relevance is the fact that the RCM approach to **85** is highly modular and therefore opens up an efficient entry into congeners and analogues of tricolorin A, which may help to map the structure/activity profile of this bioactive target molecule.

5.6
(-)-Gloeosporone: Development of a Binary Metathesis Catalyst System

The "relay model" outlined in Sect. 3.3 (Scheme 17) [30b] predicts that functionalized alkenes which lead to the formation of stable 5- or 6-membered chelate carbene intermediates on exposure to **1** should constitute poor substrates for

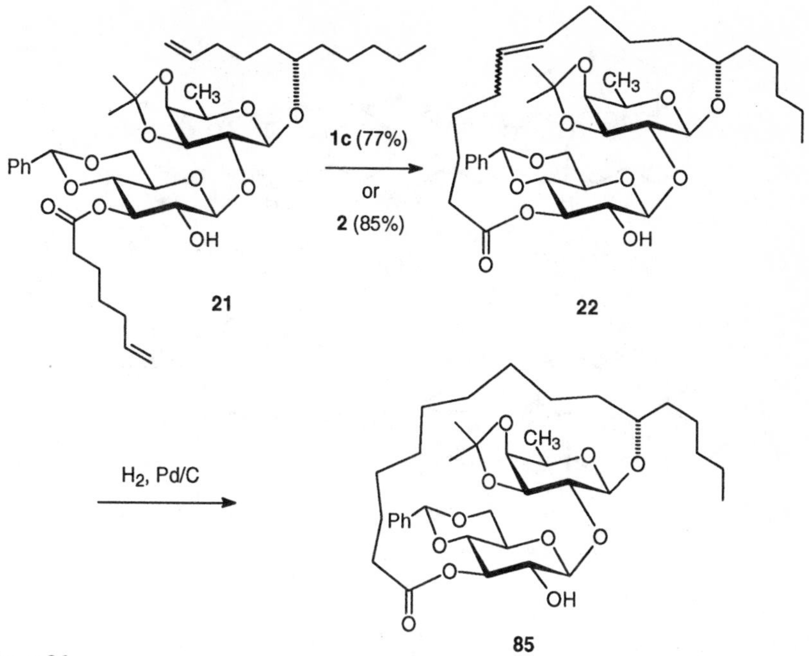

Scheme 26.

RCM-based macrocyclization reactions as they potentially sequester the catalyst in an unproductive form. A recent total synthesis of gloeosporone **86** addresses this aspect and describes a simple, yet efficient way to overcome this limitation [31].

Gloeosporone is the germination self inhibitor produced by the fungus *Colletotrichum gloeosporoides*. This compound has attracted much attention both for its interesting structural and biological properties [48]. The retrosynthetic analysis shown in Scheme 27 suggests that RCM might pave the way for an unprecedentedly short synthesis of this macrolide, provided that diene **89** can be cyclized despite the unfavorable relay which may be formed at its 4-pentenoate entity [31].

Diastereomerically pure diene **89** (R=SiMe$_2$tBu) was assembled in a straightforward manner starting from cycloheptene **90** (Scheme 28). It is not surprising to find that this particular substrate *fails* to cyclize when reacted with **1d** under standard high dilution conditions.

In order to destabilize the likely unproductive 6-membered chelate structure of type **K** (Scheme 17) that might be formed if the catalyst reacts with the 4-pentenoate entity, the cyclization was run in the presence of a Lewis acid which competes with the evolving carbene for the Lewis basic ester group. Such an additive has to be compatible with the RCM catalyst, should provoke a minimum of acid-catalyzed side reactions, and must undergo a *kinetically labile* coordination with the relay substituent. Ti(OiPr)$_4$ was found to meet these stringent requirements

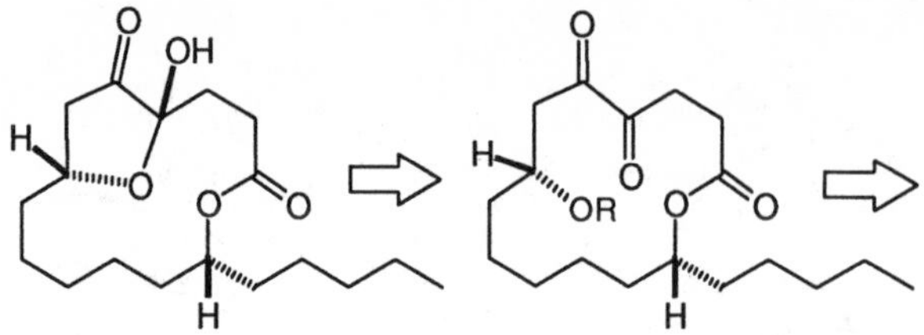

Scheme 27.

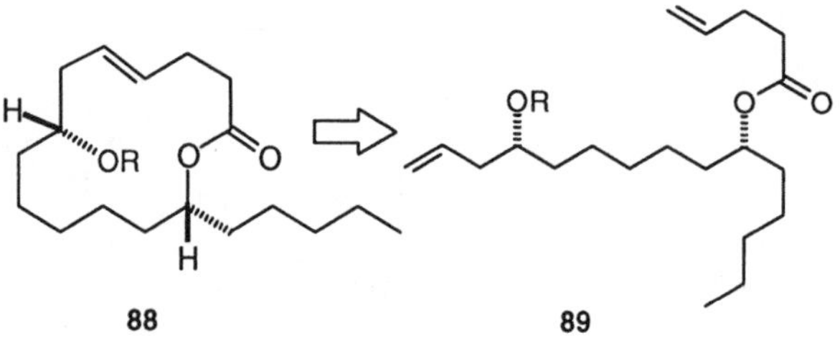

[31]. Thus, reaction of diene **89** with catalytic amounts of **1d** in the presence of sub-stoichiometric amounts of Ti(OiPr)$_4$ led to its smooth cyclization with formation of the desired 14-membered ring **96** (E:Z=2.7:1) in excellent yield. This product was then transformed into **86** by oxidation of the double bond and subsequent cleavage of the silyl protecting group. Note that this strategy merges the (E)/(Z)-mixture formed by RCM into a single diketone **97** en route to the final target.

Several other aspects of this total synthesis are also noteworthy [31]: enantiomerically pure gloeosporone has been obtained in only eight synthetic operations with an overall yield of 18% starting from cycloheptene as the substrate. This approach is much shorter and more efficient than any of the previous macrolactonization strategies reported in the literature [48]. Importantly, the number of "unproductive" protection/deprotection steps is kept to a minimum and the synthesis plan is flexible enough for making various analogues of this biologically active target. Finally it should be noted that all C-C bond formations in the sequence shown in Scheme 28 are transition metal catalyzed and lead to the concomitant formation of the chiral centers in stereomerically pure form. Therefore this example highlights the notion that a well-orchestrated interplay of RCM with other transition metal catalyzed reactions opens up flexible, modular and performant avenues to rather complex target molecules [3c, 31].

Acknowledgements: I thank my coworkers Klaus Langemann, Nicole Kindler, Thomas Müller and Günter Seidel for their intellectual and experimental contributions to our endeavors in this field. The excellent cooperation with Priv.-Doz. Dr. Walter Leitner, Mülheim, and his coworkers Daniel Koch and Christian Six

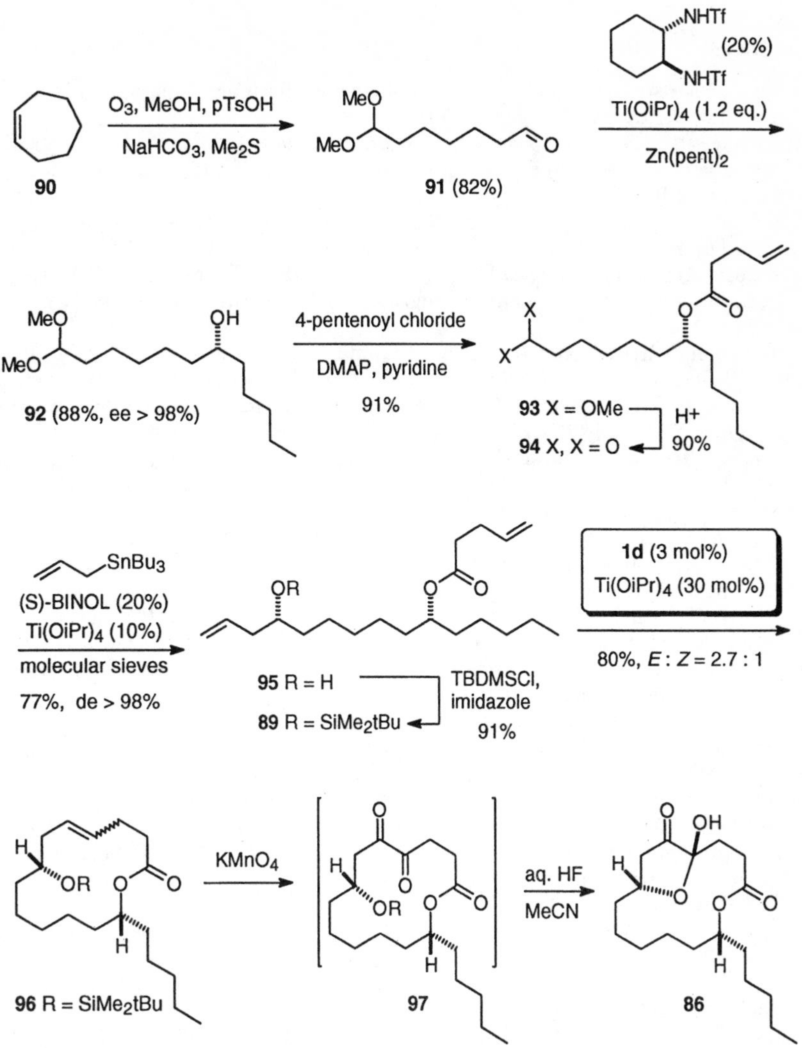

Scheme 28.

in the scCO$_2$ project, as well as with Prof. Pierre Dixneuf, Dr. Christian Bruneau and Michel Picquet, Université de Rennes, France, during the search for new metathesis catalysts is acknowledged with gratitude. We thank the Fonds der Chemischen Industrie and the Deutscher Akademischer Austauschdienst (DAAD) for financial support.

6
References

1. For a review see: Ivin KJ, Mol JC (1997) Olefin metathesis and metathesis polymerization, 2nd ed, Academic Press, New York
2. See inter alia: (a) Natta G, Dall'Asta G, Motroni G (1964) J Polym Sci Poly Lett B2:349. (b) Natta G, Dall'Asta G, Porri L (1965) Makromol Chem 81:253. (c) Michelotti FW, Keaveney WP (1965) J Polym Sci A3:895
3. For recent reviews see: (a) Grubbs RH, Miller SJ, Fu GC (1995) Acc Chem Res 28:446. (b) Schuster M, Blechert S (1997) Angew Chem 109:2124; Angew Chem Int Ed Engl 36:2036. (c) Fürstner A (1997) Topics in Catalysis 4:285. (d) Schmalz HG (1995) Angew Chem 107:198; Angew Chem Int Ed Engl 34:1833. (e) Armstrong SK (1998) J Chem Soc Perkin 1 371
4. Ruthenium vinylcarbenes: (a) Nguyen ST, Johnson LK, Grubbs RH, Ziller JW (1992) J Am Chem Soc 114:3974. (b) Nguyen ST, Grubbs, RH, Ziller JW (1993) J Am Chem Soc 115:9858. (c) Fu GC, Nguyen ST, Grubbs RH (1993) J Am Chem Soc 115:9856. (d) Wu Z, Nguyen ST, Grubbs RH, Ziller JW (1995) J Am Chem Soc 117:5503
5. Other ruthenium carbenes: (a) Schwab P, France MB, Ziller JW, Grubbs RH (1995) Angew Chem 107:2179; Angew Chem Int Ed Engl 34:2039. (b) Schwab P, Grubbs RH, Ziller JW (1996) J Am Chem Soc 118:100
6. Fürstner A, Picquet M, Bruneau C, Dixneuf PH (1998) Chem. Commun. 1315
7. Fürstner A, Koch D, Langemann K, Leitner W, Six C (1997) Angew Chem 109:2562; Angew Chem Int Ed Engl 36:2466
8. This type of rearrangement has been described previously: Binger P, Müller P, Benn R, Mynott R (1989) Angew Chem 101:647; Angew Chem Int Ed Engl 28:610
9. Wilhelm TE, Belderrain TR, Brown SN, Grubbs RH (1997) Organometallics 16:3867
10. Belderrain, TR, Grubbs RH (1997) Organometallics 16:4001
11. Dias EL, Nguyen ST, Grubbs RH (1997) J Am Chem Soc 119:3887
12. Mohr B, Lynn DM, Grubbs RH (1996) Organometallics 15:4317
13. Hérisson JL, Chauvin Y (1970) Makromol Chem 141:161
14. Tallarico JA, Bonitatebus PJ, Snapper ML (1997) J Am Chem Soc 119:7157
15. (a) Stumpf AW, Saive E, Demonceau A, Noels AF (1995) J Chem Soc Chem Commun 1127. (b) Demonceau A, Stumpf AW, Saive E, Noels AF (1997) Macromolecules 30:3127
16. Hafner A, Mühlebach A, van der Schaaf PA (1997) Angew Chem 109:2213; Angew Chem Int Ed Engl 36:2121
17. (a) Herrmann WA, Schattenmann WC, Nuyken O, Glander SC (1996) Angew Chem 108:1169; Angew Chem Int Ed Engl 35:1087. (b) Hiraki K, Kuroiwa A, Hirai H (1971) J Polym Sci A-1 9:2323. (c) Porri L, Rossi R, Diversi P, Lucherini A (1974) Makromol Chem 175:3097
18. Schrock RR, Murdzek, JS, Bazan GC, Robbins J, DiMare M, O'Regan M (1990) J Am Chem Soc 112:3875
19. (a) Kirkland TA, Grubbs RH (1997) J Org Chem 62:7310. (b) For RCM of non-terminal olefins with catalyst 24 see also: Fu GC, Grubbs RH (1992) J Am Chem Soc 114:7324. (c) Fu GC, Grubbs RH (1992) J Am Chem Soc 114:5426
20. (a) Kim SH, Bowden N, Grubbs RH (1994) J Am Chem Soc 116:10801. (b) Zuercher WJ, Hashimoto M, Grubbs RH (1996) J Am Chem Soc 118:6634
21. Huwe CM, Velder J, Blechert S (1996) Angew Chem 108:2542; Angew Chem Int Ed Engl 35:2376
22. Overkleeft HS, Pandit UK (1996) Tetrahedron Lett 37:547
23. (a) Fürstner A, Nikolakis K, unpublished results. (b) Nikolakis K (1998) PhD Thesis, University of Dortmund
24. (a) Rutjes FPJT, Schoemaker HE (1997) Tetrahedron Lett 38:677. (b) Schuster M, Pernerstorfer J, Blechert S (1996) Angew Chem 108:2111; Angew Chem Int Ed Engl 35:1979
25. Arisawa M, Takezawa E, Nishida A, Mori M, Nakagawa M (1997) Synlett 1179

26. For a pertinent example see: Fujimura O, Fu GC, Rothemund PWK, Grubbs RH (1995) J Am Chem Soc 117:2355
27. Diver ST, Schreiber SL (1997) J Am Chem Soc 119:5106
28. (a) Armstrong SK, Christie BA (1996) Tetrahedron Lett 37:9373. (b) Shon YS, Lee TR (1997) Tetrahedron Lett 38:1283
29. For leading references on the cyclization of medium sized rings with either 1 or 24 as catalysts see: (a) Fürstner A, Müller T (1997) Synlett 1010. (b) Fürstner A, Langemann K (1996) J Org Chem 61:8746. (c) Martin SF, Liao Y, Chen HJ, Pätzel M, Ramser MN (1994) Tetrahedron Lett 35:6005. (d) Martin SF, Liao Y, Wong Y, Rein T (1994) Tetrahedron Lett 35:691. (e) Winkler JD, Stelmach JE, Axten J (1996) Tetrahedron Lett 37:4317. (f) Visser MS, Heron NM, Didiuk MT, Sagal JF, Hoveyda AH (1996) J Am Chem Soc 118:4291. (g) Clark JS, Kettle JG (1997) Tetrahedron Lett 38:123, 127. (h) Miller SJ, Grubbs RH (1995) J Am Chem Soc 117:5855. (i) Miller SJ, Kim SH, Chen ZR, Grubbs RH (1995) J Am Chem Soc 117:2108. (j) Linderman RJ, Siedlecki J, O'Neill SA, Sun H (1997) J Am Chem Soc 119:6919. (k) Crimmins MT, Choy AL (1997) J Org Chem 62:7548. (l) Joe D, Overman LE (1997) Tetrahedron Lett 38:8635. (m) Chang S, Grubbs RH (1997) Tetrahedron Lett 38:4757. (n) Barrett AGM, Baugh SPD, Gibson VC, Giles MR, Marshall EL, Procopiou PA (1996) J Chem Soc Chem Commun 2231
30. (a) Fürstner A, Langemann K (1996) J Org Chem 61:3942. (b) Fürstner A, Langemann K (1997) Synthesis 792. (c) Fürstner A (1997) In: Helmchen G (ed) Organic Synthesis via Organometallics OSM5. Vieweg, Braunschweig, p 309
31. Fürstner A, Langemann K (1997) J Am Chem Soc 119:9130
32. Fürstner A, Müller T (1998) J Org Chem 63:424
33. Fürstner A, Kindler N (1996) Tetrahedron Lett 37:7005
34. (a) Borer BC, Deerenberg S, Bieräugel H, Pandit UK (1994) Tetrahedron Lett 35:3191. (b) Magnier E, Langlois Y (1998) Tetrahedron Lett 39:837
35. (a) Nicolaou KC, He Y, Vourloumis D, Vallberg H, Yang Z (1996) Angew Chem 108:2554; Angew Chem Int Ed Engl 35:2399. (b) Yang Z, He Y, Vourloumis D, Vallberg H, Nicolaou KC (1997) Angew Chem 109:170; Angew Chem Int Ed Engl 36:166. (c) Meng D, Su DS, Balog A, Bertinato P, Sorensen EJ, Danishefsky SJ, Zheng YH, Chou TC, He L, Horwitz SB (1997) J Am Chem Soc 119:2733. (d) Bertinato P, Sorensen EJ, Meng D, Danishefsky SJ (1996) J Org Chem 61:8000. (e) Schinzer D, Limberg A, Bauer A, Böhm OM, Cordes M (1997) Angew Chem 109:543; Angew Chem Int Ed Engl 36:523. (f) Nicolaou KC, He Y, Roschangar F, King NP, Vourloumis D, Li T (1998) Angew Chem 110:89; Angew Chem Int Ed Engl 37:84. (g) Nicolaou KC, He Y, Vourloumis D, Vallberg H, Roschangar F, Sarabia F, Ninkovic S, Yang Z, Trujillo JI (1997) J Am Chem Soc 119:7960. (h) Meng D, Bertinato P, Balog A, Su DS, Kamenecka T, Sorensen EJ, Danishefsky SJ (1997) J Am Chem Soc 119:10073
36. (a) Marsella MJ, Maynard HD, Grubbs RH (1997) Angew Chem 109:1147; Angew Chem Int Ed Engl 36:1101. (b) König B, Horn C (1996) Synlett 1013. (c) Delgado M, Martin JD (1997) Tetrahedron Lett 38:8387
37. (a) Miller SJ, Blackwell HE, Grubbs RH (1996) J Am Chem Soc 118:9606. (b) Clark TD, Ghadiri MR (1995) J Am Chem Soc 117:12364
38. (a) Mohr B, Weck M, Sauvage JP, Grubbs RH (1997) Angew Chem 109:1365; Angew Chem Int Ed Engl 36:1308. (b) Dietrich-Buchecker C, Rapenne G, Sauvage JP (1997) J Chem Soc Chem Commun 2053
39. McKervey MA, Pitarch M (1996) J Chem Soc Chem Commun 1689
40. Kim SH, Figueroa I, Fuchs PL (1997) Tetrahedron Lett 38:2601
41. For an alternative approach see: (a) Fürstner A, Weintritt H (1997) J Am Chem Soc 119:2944. (b) Fürstner A, Weintritt H (1998) J Am Chem Soc 120:2817
42. (a) Harrity JPA, Visser MS, Gleason JD, Hoveyda AH (1997) J Am Chem Soc 119:1488. (b) Morken, JP, Didiuk, MT, Visser MS, Hoveyda AH (1994) J Am Chem Soc 116:3123. (c) Heron NM, Adams JA, Hoveyda AH (1997) J Am Chem Soc 119:6205. (d) Visser MS, Heron NM, Didiuk MT, Sagal JF, Hoveyda AH (1996) J Am Chem Soc 118:4291

43. Peters JU, Blechert S (1997) J Chem Soc Chem Commun 1983
44. (a) Meyer C, Cossy J (1997) Tetrahedron Lett 38:7861. (b) Cassidy JH, Marsden SP, Stemp G (1997) Synlett 1411
45. For leading references see: (a) Nicolaou KC, Winssinger N, Pastor J, Ninkovic S, Sarabia F, He Y, Vourloumis D, Yang Z, Li T, Giannakakou P, Hamel E (1997) Nature 387:268. (b) Nicolaou KC, Vourloumis D, Li T, Pastor J, Winssinger N, He Y, Ninkovic S, Sarabia F, Vallberg H, Roschangar F, King NP, Finlay, MRV, Giannakakou P, Verdier-Pinard P, Hamel E (1997) Angew Chem 109:2181; Angew Chem Int Ed Engl 36:2097. (c) Piscopio AD, Miller JF, Koch K (1997) Tetrahedron Lett 38:7143. (d) Pernerstorfer J, Schuster M, Blechert S (1997) J Chem Soc Chem Commun 1949. (e) Schuster M, Lucas N, Blechert S (1997) J Chem Soc Chem Commun 823. (f) Peters JU, Blechert S (1997) Synlett 348. (g) Boger DL, Chai W, Ozer RS, Andersson CM (1997) Bioorg Med Chem Lett 7:463. (h) van Maarseveen JH, den Hartog JAJ, Engelen V, Finner E, Visser G, Kruse CG (1996) Tetrahedron Lett 37:8249
46. Crimmins MT, King BW (1996) J Org Chem 61:4192
47. For syntheses of tricolorin A based on macrolactonization reactions see: (a) Larson DP, Heathcock CH (1997) J Org Chem 62:8406. (b) Lu SF, O'yang QQ, Guo ZW, Yu B, Hui YZ (1997) J Org Chem 62:8400
48. For syntheses of gloeosporone by macrolactonization strategies see: (a) Adam G, Zibuck R, Seebach D (1987) J Am Chem Soc 109:6176. (b) Schreiber SL, Kelly SE, Porco JA, Sammakia T, Suh EM (1988) J Am Chem Soc 110:6210. (c) Curtis NR, Holmes AB, Looney MG, Pearson ND, Slim GC (1991) Tetrahedron Lett 32:537. (d) Takano S, Shimazaki Y, Takahashi M, Ogasawara K (1988) J Chem Soc Chem Commun 1004. (e) Matsushita M, Yoshida M, Zhang Y, Miyashita M, Irie H, Ueno T, Tsurushima T (1992) Chem Pharm Bull 40:524
49. Hammer K, Undheim K (1997) Tetrahedron 53:5925
50. Dyatkin A (1997) Tetrahedron Lett 38:2065
51. Schneider MF, Junga H, Blechert S (1995) Tetrahedron 51:13003
52. Maier ME, Langenbacher D, Rebien F (1995) Liebigs Ann 1843
53. Morehead A, Grubbs RH (1998) J Chem Soc Chem Commun 275
54. Oishi T, Nagumo Y, Hirama M (1997) Synlett 980
55. Garro-Hélion F, Guibé F (1996) J Chem Soc Chem Commun 641
56. Mascarenas JL, Rumbo A, Castedo L (1997) J Org Chem 62:8620
57. Leeuwenburgh MA, Overkleeft HS, van der Marel GA, van Boom JH (1997) Synlett 1263
58. Barrett AGM, Baugh SPD, Gibson VC, Giles MR, Marshall EL, Procopiou PA (1997) J Chem Soc Chem Commun 155
59. Delgado M, Martin JD (1997) Tetrahedron Lett 38:6299
60. Note added in proof: For an improved procedure for the preparation of Grubbs carbenes see: Wolf J, Stüer W, Grünwald C, Werner H, Schwab P, Schulz M (1998) Angew Chem 110:1165; Angew Chem Int Ed Engl 37:1124
61. For an indirect, but stereoselective approach to (z)- comfigurated cycloalkenes via ring-closing-metathesis of diynes, followed by semi-reduction of the macrocyclic cycloalkynes thus formed see: Fürstner A, Seidel G (1998) Angew Chem 110:1758; Angew Chem Int Ed Engl 37:1734

Ring-Closing Metathesis in the Synthesis of Epothilones and Polyether Natural Products

K.C. Nicolaou, N. Paul King, and Yun He

The increasing popularity of ring-closing metathesis (RCM) can be attributed to the development of transition metal complexes **1, 2** and **3** as initiators. These compounds efficiently promote the RCM process, are compatible with a wide range of chemical functionalities and can be used without recourse to rigorously controlled reaction conditions. In this chapter, applications of this technology to the preparation of the 16-membered macrolactone core of the epothilones culminating in the total synthesis of epothilones A, B and E will be presented. The preparation of a diverse array of analogs using both solution and solid-phase techniques will also be discussed. The use of the cyclopentadienyl titanium complexes **93** and **110** for olefination/olefin metathesis will also be described, with particular emphasis on their potential in the synthesis of polyether natural products.

Keywords: Ring-closing metathesis, Epothilones, Tebbe reagent, Petasis reagent, Maitotoxin

1
Introduction

Ring-closing metathesis (RCM) is an extremely powerful method for transforming acyclic dienes into unsaturated cyclic systems [1]. The process is believed to proceed via the catalytic cycle outlined in Scheme 1, and the principal driving force for the reaction is the gain in entropy resulting from the loss of ethene.

Although RCM technology has been known for over 15 years, early examples utilized poorly defined and inefficient catalyst systems which exhibited limited functional group tolerance [2]. These factors made the process unsuitable for most synthetic applications. The recent explosive growth in the area can be attributed primarily to the work of Schrock [3] and Grubbs [4], who developed the stable, well-defined and efficient metathesis initiators 1, 2 and 3 (Fig. 1).

These complexes generate the active catalytic species in situ and are now used routinely for the preparation of 5-, 6- and 7-membered carbo- and hetero-cyclic ring systems from the appropriate diene precursors. Contrary to original expectations, the preparation of medium and large ring systems has also been demonstrated, even in cases where the diene substrate is not predisposed to cyclization [5]. In addition to their high activity, 1, 2 and 3 exhibit exceptional tolerance to a wide variety of functional groups. In the context of complex molecule construction, this can allow RCM to play a pivotal role in the development of concise and efficient synthetic routes in which protecting group transformations are kept to a minimum. Furthermore, the experimental ease with which the reac-

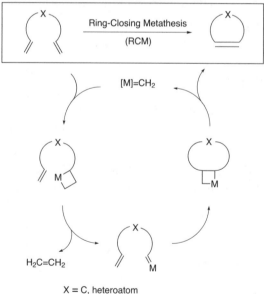

X = C, heteroatom
[M]=CH$_2$ = Transition metal alkylidene

Scheme 1. Mechanism for ring-closing metathesis

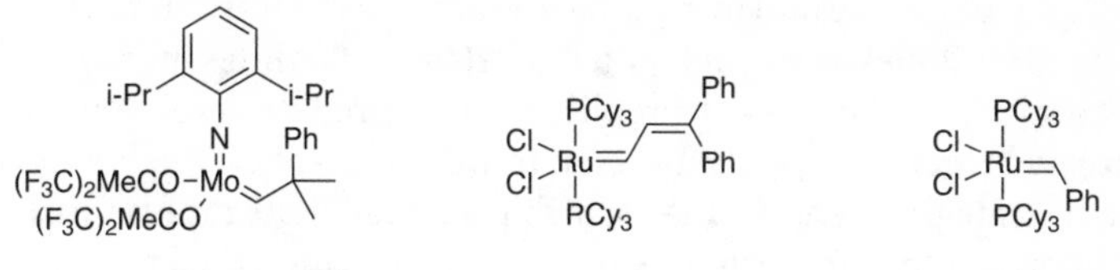

Fig. 1. Commonly used metathesis initiators

tions can be performed, and the possibility of manipulating the olefinic functionality formed by the process, has made RCM an increasingly powerful reaction for use in modern organic synthesis.

This article will outline firstly the application of RCM technology to the preparation of the epothilones with particular emphasis on the generality and mild nature of the process. The second section will describe the use of cyclopentadienyl titanium reagents in metathesis processes and, in particular, their application to the preparation of polyether segments of marine neurotoxins.

2
Total Synthesis of Epothilones Via Ring-Closing Metathesis

2.1
Introduction

Epothilones A, B and E (**4, 5** and **6**) (Fig. 2) are representative members of a new class of bacterially derived natural products which exhibit potent biological activity. Isolated by Höfle and coworkers [6] from a soil sample collected near the Zambesi river, the compounds have provided a great deal of excitement in the scientific community due to their potent cytotoxicity against a number of multiple drug-resistant tumor cell lines and because of the mechanism by which they exert this effect. Like Taxol [7], the epothilones promote the combination of α- and β-tubulin subunits and stabilize the resulting microtubule structures. This mode of action inhibits the cell division process and is, therefore, an attractive strategy for cancer chemotherapy [7, 8].

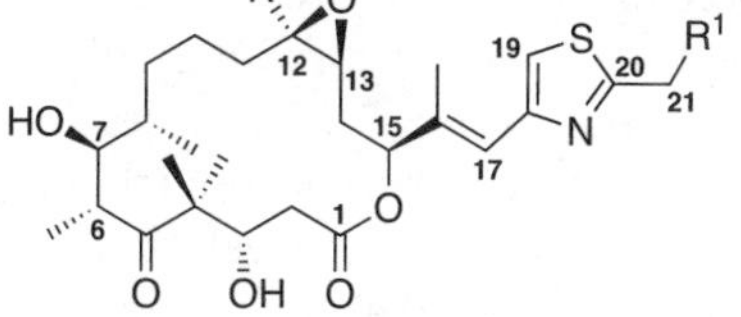

4: R = R^1 = H, epothilone A
5: R = Me, R^1 = H, epothilone B
6: R = H, R^1 = OH, epothilone E

Fig. 2. Structures and numbering of the epothilones

In addition to the rare and intriguing mechanism of antitumor activity exhibited by the epothilones, their high potency (2000–3000 times more than Taxol in some assays [8a]) and chemical simplicity (compared to Taxol) has made them high priority targets for total synthesis. In this connection, highly flexible and convergent strategies were required which would facilitate rapid construction of the epothilones. More importantly, this would provide efficient access to a library of diverse analogs for biological screening purposes with the ultimate goal of fine-tuning the pharmacological properties of the natural molecules.

2.2
Total Synthesis of Epothilones A, B and E

Prior to the synthesis of the epothilones, the use of RCM for the preparation of macrolactones had received little attention [5]. In part, this could be attributed to the generally held assumption that only conformational biased precursors would undergo cyclization. Although several examples of macro-RCM had been reported, notably by Pandit [9], Hoveyda [10] and Fürstner [11], preparation of the densely functionalized 16-membered lactone core of the epothilones was not a trivial undertaking. Preliminary studies focused on the preparation and cyclization of model substrates in order to assess the viability of the RCM approach and to provide precedent for subsequent, more ambitious, synthetic endeavors.

2.2.1
Model Studies

Nicolaou et al. were the first to report the successful use of RCM to prepare the 16-membered macrolactone nucleus of the epothilones and present a strategy for their total synthesis based on this reaction. The approach involved formation of the C12,C13 olefin and is outlined in Scheme 2 [12, 13].

The RCM precursor 11 was constructed in a highly convergent manner from the three fragments 7, 8 and 10 via sequential esterification and aldol reactions. Treatment of triene 11 with ruthenium initiator 3 (0.1 equiv) at 25°C in dichloromethane (0.003 M) resulted in the formation of a single macrolactone product 13 in 85% yield. Under similar conditions, the C6,C7 diastereomer, 12, underwent smooth cyclization to give the corresponding 6S,7R macrolactone (not shown). Despite lacking the C3 hydroxyl functionality and the thiazole moiety present in the natural compounds (Fig. 2), the successful transformation of 11→ 13 indicated the viability of the metathesis approach for the formation of the 16-membered macrolactone and led to the development of a second generation model which incorporated the desired aromatic substituent [13] (Scheme 3).

The tetraene precursor 14, assembled in a similar way to 11, underwent smooth cyclization using the ruthenium initiator 3 (0.1 equiv) to give macrolactone 15, again in good yield and with complete E-selectivity. Despite the incorrect olefin geometry, transformation into epoxides 16 provided further encour-

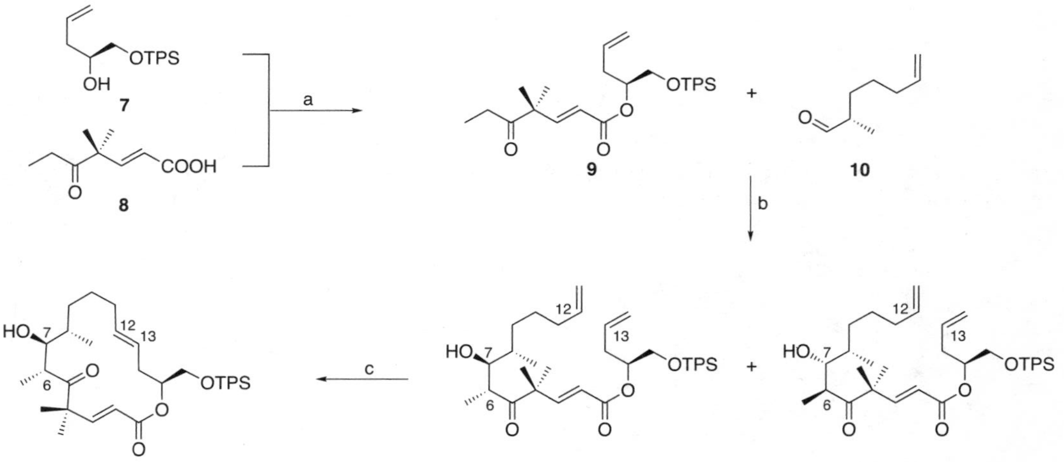

Scheme 2. Synthesis of the macrolactone core of epothilone A via RCM. (*a*) EDC, DMAP, CH$_2$Cl$_2$, 86%; (*b*) LDA, THF, −78°C, 75% (**11:12**=4:3); (*c*) 0.1 equiv of RuCl$_2$(=CHPh)(PCy$_3$)$_2$ (**3**), CH$_2$Cl$_2$, 0.003 M, 25°C, 12 h, 85% (Nicolaou et al.) [12, 13]

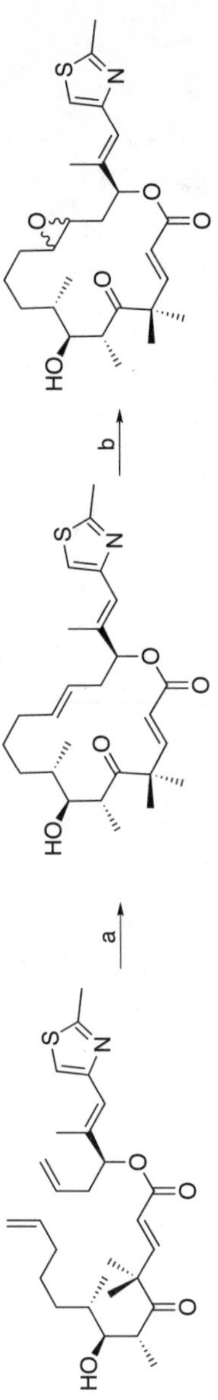

14 15 **16a**: β-epoxide **16b**: α-epoxide

Scheme 3. Metathesis in the presence of thiazole and epoxidation of the resulting triene. (*a*) 0.1 equiv of $RuCl_2(=CHPh)(PCy_3)_2$ (3), CH_2Cl_2, 0.001 M, 25°C, 12 h, 66%; (*b*) methyltrifluoromethyldioxirane, CH_3CN, 0°C, 75% (**16a**:**16b**=4:1) (Nicolaou et al.) [13]

agement for the extension of this approach to the total synthesis of the epothilones.

In parallel investigations, Danishefsky and coworkers accomplished the preparation of the 16-membered lactone of a model epothilone system via an alternative C9,C10 disconnection [14] (Scheme 4). In this case, coupling of epoxy-alcohol 17 with acids 18a and 18b afforded trienes 19a and 19b respectively. RCM of 19a under the influence of ruthenium initiator 3 produced dienes 20a as a 1:1 mixture of Z:E-isomers. Under identical conditions, cyclization of 19b produced a single product 20b (tentatively assigned as the Z-isomer). The variable stereoselectivity observed in these reactions was inconsequential since the olefinic functionality could be reduced to afford the corresponding saturated macrolactones. Schrock's molybdenum initiator 1 promoted the cyclization of 19a and 19b with similar efficacy [14].

Despite comprehensive studies, extension of this C9,C10 strategy to the preparation of a fully functionalized epothilone intermediate proved unattainable, demonstrating limitations of the RCM process [14].

2.2.2
Total Synthesis of Epothilone A

The C12,C13 disconnection strategy has been employed by several groups to complete total syntheses of epothilone A (4). Nicolaou et al. were the first to report the successful application of this strategy to the construction of a naturally occurring epothilone [13, 15] (Scheme 5).

Aldol reaction of keto-acid 21 with aldehyde 10 and esterification of the resulting acids with alcohol 22 led rapidly to cyclization precursor 23 and its 6S,7R-diastereomer (not shown). RCM using ruthenium initiator 3 (0.1 equiv) in dichloromethane (0.0015 M) at 25°C afforded macrolactones 24a and 24b in a 1.2:1 ratio. Deprotection and epoxidation of the desired macrolactone, 24a, afforded epothilone A (4) via 25a (epothilone C) (Scheme 5). Varying a number of reaction parameters, such as solvent, temperature and concentration, failed to improve significantly the Z-selectivity of the RCM. However, in the context of the epothilone project, the formation of the E-isomer 24b could actually be viewed as beneficial since it allowed preparation of the epothilone A analog 26 for biological evaluation.

Using a similar C12,C13 disconnection approach, Schinzer et al. also achieved a total synthesis of epothilone A (4) [16]. The key step involved a highly selective aldol reaction between ketone 27 and aldehyde 10 to afford exclusively alcohol 28 with the correct C6,C7 stereochemistry (Scheme 6). Further elaboration led to triene 29, which underwent RCM using ruthenium initiator 3 in dichloromethane at 25°C, to afford macrocyles 30 in high yield (94%). Although no selectivity was observed (Z:E=1:1), deprotection and epoxidation of the desired Z-isomer (30a) completed the total synthesis [16].

Following the completion of their first synthesis of epothilone A (4) based on a macroaldolisation strategy [17], Danishefsky and coworkers applied a C12,C13

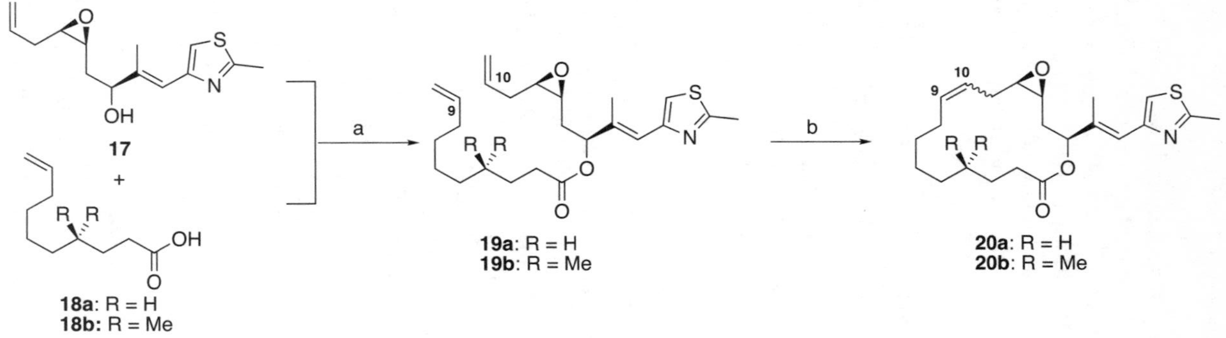

Scheme 4. Metathesis in the presence of an epoxide and thiazole. (*a*) EDC, DMAP, CH$_2$Cl$_2$, 89–91%, (*b*) 0.1 equiv of RuCl$_2$(= CHPh)(PCy$_3$)$_2$ (3), C$_6$H$_6$, 0.005 M, 25°C, 24 h, **20a** (45%, Z:E=1:1); **20b** (70%, Z:E=1:0) (Danishefsky et al.) [14]

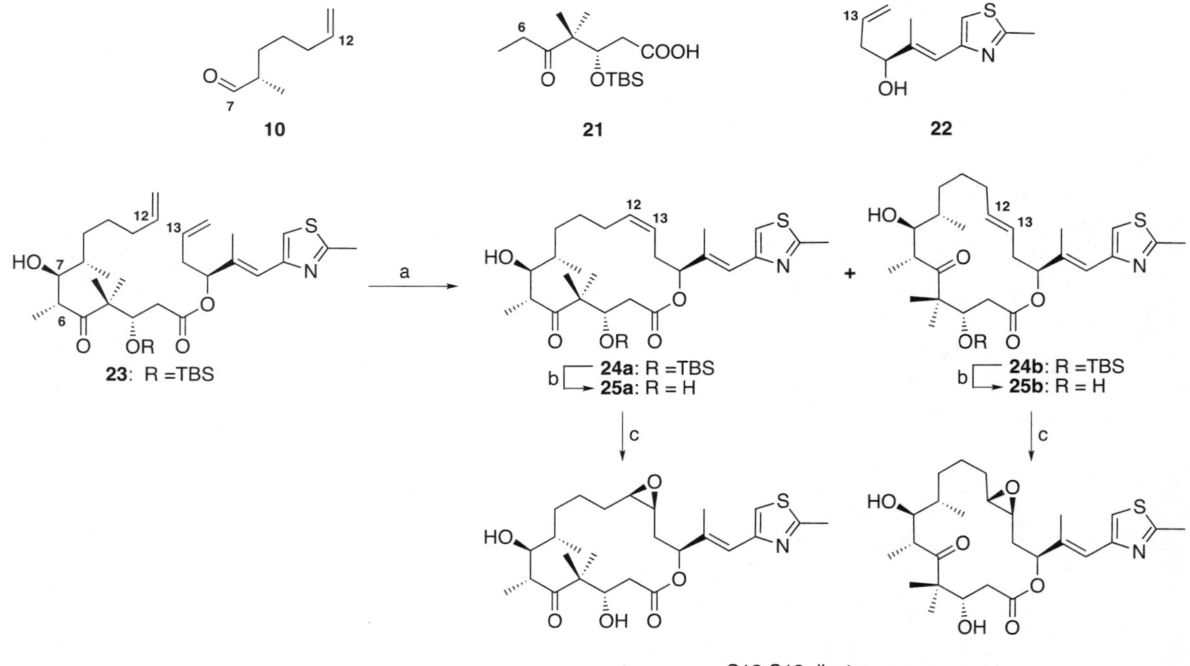

Scheme 5. Total synthesis of epothilone A (**4**) and analogs. (*a*) 0.1 equiv of RuCl$_2$(=CHPh)(PCy$_3$)$_2$ (**3**), CH$_2$Cl$_2$, 0.0015 M, 25°C, 20 h, 85% (**24a**:**24b**=1.2:1); (*b*) TFA, CH$_2$Cl$_2$, 90–92%;(*c*) methyltrifluoromethyldioxirane, 50–70% (Nicolaou et al.) [13, 15]

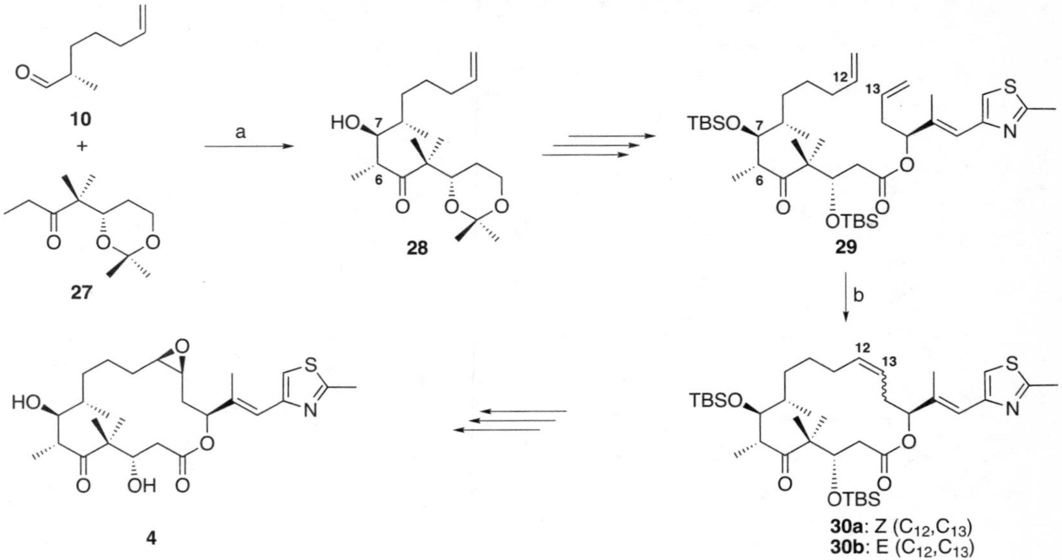

Scheme 6. Total synthesis of epothilone A (4). (a) LDA, THF, −78 °C, 70%; (b) RuCl$_2$(=CHPh)(PCy$_3$)$_2$ (3), CH$_2$Cl$_2$, 25°C, 12 h, 94% (**30a:30b**=1:1) (Schinzer et al.) [16]

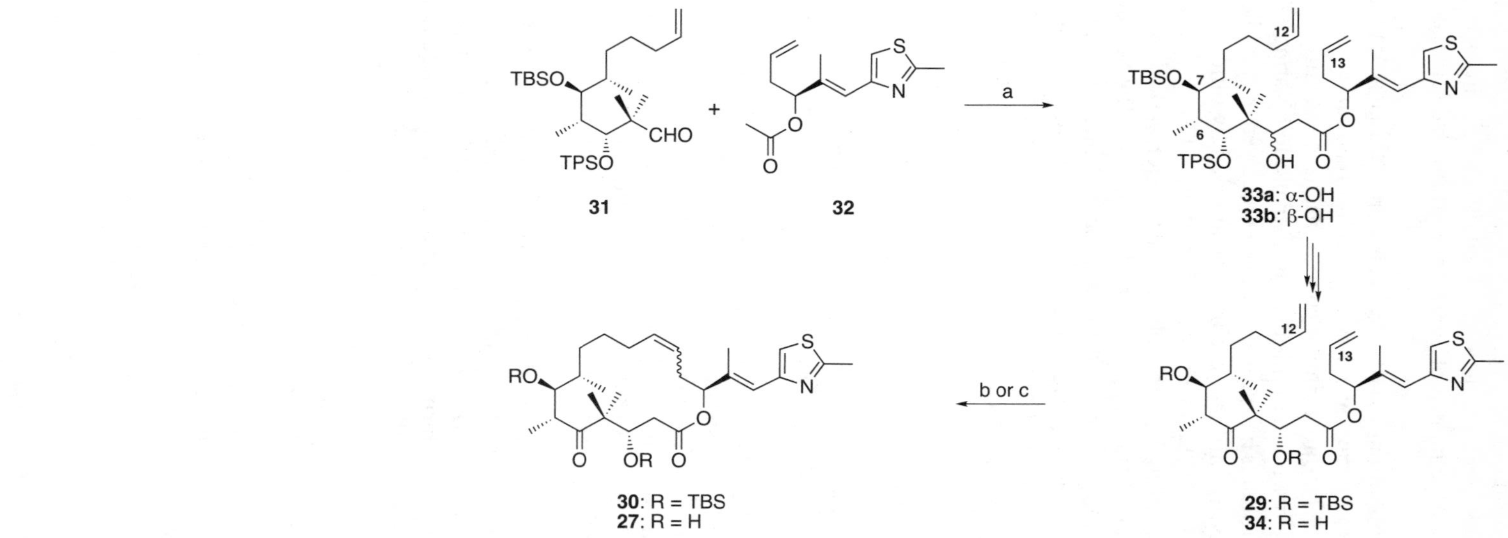

Scheme 7. Synthesis of epothilone A precursor (27) via RCM. (*a*) LDA, THF, −78°C, 70% (33a:33b=1:1); (*b*) 0.5 equiv of RuCl₂{=CHPh}(PCy₃)₂ (3), C₆H₆, 0.001 M, 25°C, 4 h, 86% (30, Z:E=1.7:1), 65% (27, Z:E= 1:2); (*c*) 0.2 equiv of Mo(=CHCMe₂Ph){N[2,6-(*i*-Pr)₂C₆H₃]}[OCMe(CF₃)₂]₂ (1), C₆H₆, 0.001 M, 25°C, 1 h, 86% (30, Z:E=1:2) (Danishefsky et al.) [14b, 17b]

disconnection strategy to prepare a range of epothilone A intermediates [14b, 17b]. Intermolecular aldol reaction between acetate **32** and aldehyde **31** afforded trienes **33** (Scheme 7). A sequence of functional group manipulations led to triene **29** and **34**.

RCM of diene **29** using initiator **3** (0.5 equiv) in benzene (0.001 M) at 25°C afforded macrolactones **30** in 86% yield, with a slight preponderance of the desired Z-isomer (Z:E=1.7:1). Under similar conditions, triene **34** afforded macrolactones **27** in 65% yield with the undesired isomer predominating (Z:E=1:2). The Z-macrolactone products had previously been processed to epothilone A (**4**) [14b, 17a]. The selectivities of these cyclizations are slightly different to those achieved by the Nicolaou [18] and Schinzer [16] groups. The choice of solvent (dichloromethane vs. benzene) is probably responsible for these differences.

Danishefsky and coworkers also performed RCM of **29** using Schrock's molybdenum initiator **1** [14b]. In this case, the yield was identical to that obtained using **3** (86%) but the selectivity was reversed (Z:E=1:2). Additionally, trienes **33a** and **33b** were subjected to RCM to give epothilone A precursors **35** and **36** respectively [14b, 17b] (see Table 1).

2.2.3
Total Synthesis of Epothilone B

With the C12,C13 disconnection producing an effective solution to the synthesis of epothilone A (**4**), it would seem likely that the metathesis approach could be extended readily to the preparation of epothilone B (**5**). However, installation of the desired C12 methyl group requires ring-closure of a diene precursor in which one of the olefins is disubstituted. Recently, such reactions have been shown to be problematic for Grubbs' initiator **3** but more successful with Schrock's molybdenum initiator **1** [19]. Consistent with these reports, Danishefsky demonstrated that triene **38** would not undergo RCM with **3**, whereas **1** was effective in promoting the transformation of **38** into a 1:1 mixture of **39a** and **39b** in good yield [14b] (Scheme 8).

Subsequent deprotection of the desired z-isomer (**39a**) afforded desoxyepothilone B (epothilone D), which had previously been epoxidized to epothilone B (**5**) [14b]. Alcohol **37** did not undergo RCM using **1** or **3** [14b]. The failure of initiator (**1**) to effect this transformation may be due to its incompatibility with unprotected hydroxyl groups [19].

2.2.4
Total Synthesis of Epothilone E and Analogs

With the synthesis of epothilones A and B secured, subsequent studies concentrated on the preparation of analogs of the natural molecules. In addition to providing structure-activity relationships, it was anticipated that these studies would provide a further test for the generality of the RCM process. In this context, a general strategy was developed by Nicolaou et al. [20] to investigate the heterocyclic portion of the epothilone framework. In this approach the potentially sensitive vinyl iodide **40** was prepared in a highly convergent manner, similar to that previously described (Sect. 2.2.2). Cyclization using ruthenium initiator **3** (0.1 equiv) afforded macrolactones **41a** and **41b** in a 1.8:1 ratio (Scheme 9). Silyl ether deprotection liberated diols **42a** and **42b**, which were individually coupled to a variety of aromatic stannanes (R^1_3Sn-Aro), under palladium catalysis, to provide epothilone analogs **43** and **44**. Epoxidation of the hydroxymethyl thiazole analog [**43**; Aro=4-(2-hydroxymethylthiazole); (epothilone F)] afforded the naturally occurring epothilone E (**6**) [6b, 20].

Further studies by Nicolaou, Danishefsky and others have shown the C12,C13 RCM approach to be a highly reliable method of preparing the epothilone core (Table 1). The yields are consistently high but the Z:E-selectivities are variable and, as yet, unpredictable. However, close inspection of the results presented in Table 1 (and Sects. 2.2.2–2.2.4) reveals that several trends have emerged. Firstly, changing the C3 stereochemistry, from R- to the desired S-configuration, increases the amount of the desired Z-macrolactone (compare **49:23**, **69:67**, **33b:33a**, **73:29**). Greater Z-selectivity is also obtained with substrates possessing the desired 6R,7S stereochemistry compared to those with the alternative 6S,7R configuration (compare **23:47**, **34:45**, **55:57**, **59:61**). It is also pertinent to note that the rate of metathesis is highly substrate-dependent. For example, metathesis of the C4-cyclopropyl derivative **55** [21] (Table 1) was complete in 2 h, whereas the corresponding epothilone A precursor **23** (Scheme 5), possessing the C4-gem-dimethyl substituent, required 20 h to reach completion. It is thought that the cyclopropyl substituent allows **55** to adopt a reactive conformation more readily than its gem-dimethyl counterpart **23**.

2.3
Solid-Phase Synthesis of Epothilone A and Analogs

The past decade has witnessed enormous advances in the area of solid-phase combinatorial chemistry [22]. Application of this new technology to the preparation of the epothilones was highly desirable since it would assist in the preparation of a library of analogs for biological testing purposes. Although a large number of chemical reactions have undergone the transition to solid-phase processes [22b], adapting the solution-phase synthesis of the epothilones relied upon the successful transfer of the key RCM step to a solid-phase process. At the time, there was little precedent demonstrating the compatibility of initiator **3** with solid-phase techniques [23]. Additionally, two distinct solid-phase strategies could be envisaged which would also require consideration (Scheme 10).

The first approach involves RCM of a polymer bound diene **77** and results in the formation of a cyclic product **78** which remains attached to the polymer support. This product can undergo further manipulation, with cleavage from the resin occurring at a later stage. In a second, complementary approach, RCM proceeds with concomitant cleavage of the substrate from the resin (**80→81**). The latter protocol is highly attractive since only compounds possessing the correct

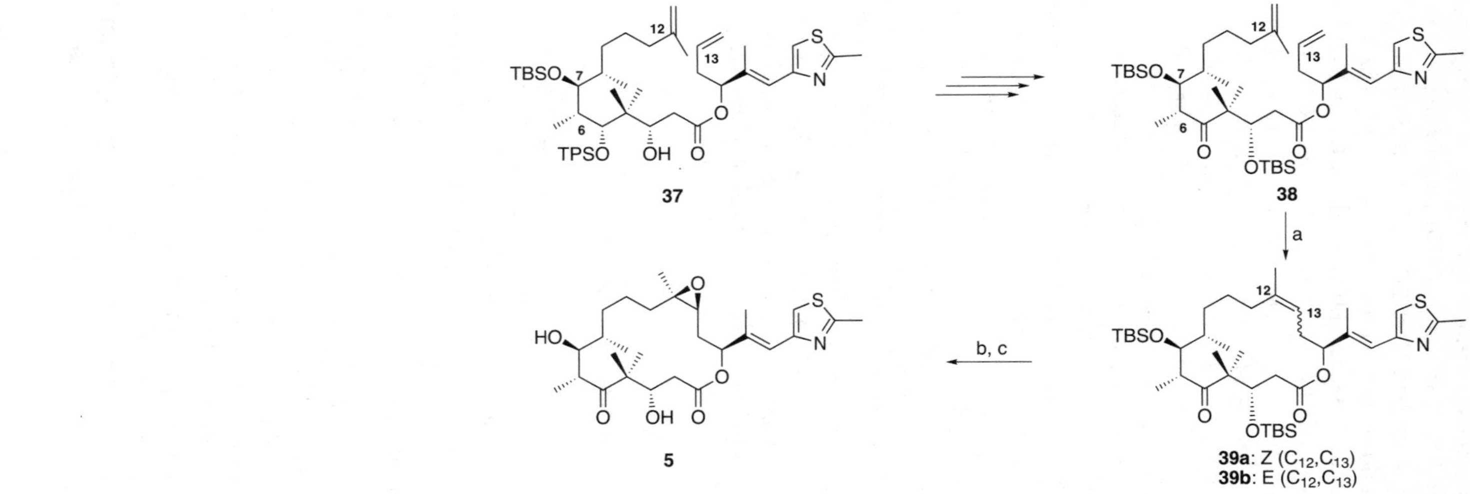

Scheme 8. Total synthesis of epothilone B (5) via RCM. (*a*) 0.2 equiv of Mo(=CHCMe₂Ph){N[2,6-(*i*-Pr)₂C₆H₃]}[OCMe(CF₃)₂]₂ (1), C₆H₆, 0.001 M, 55°C, 2 h, 86% (Z:E=1:1); (*b*) HF·Py, 94% (from **39a**); (*c*) 3,3-dimethyldioxirane, 97% (Danishefsky et al.) [14b]

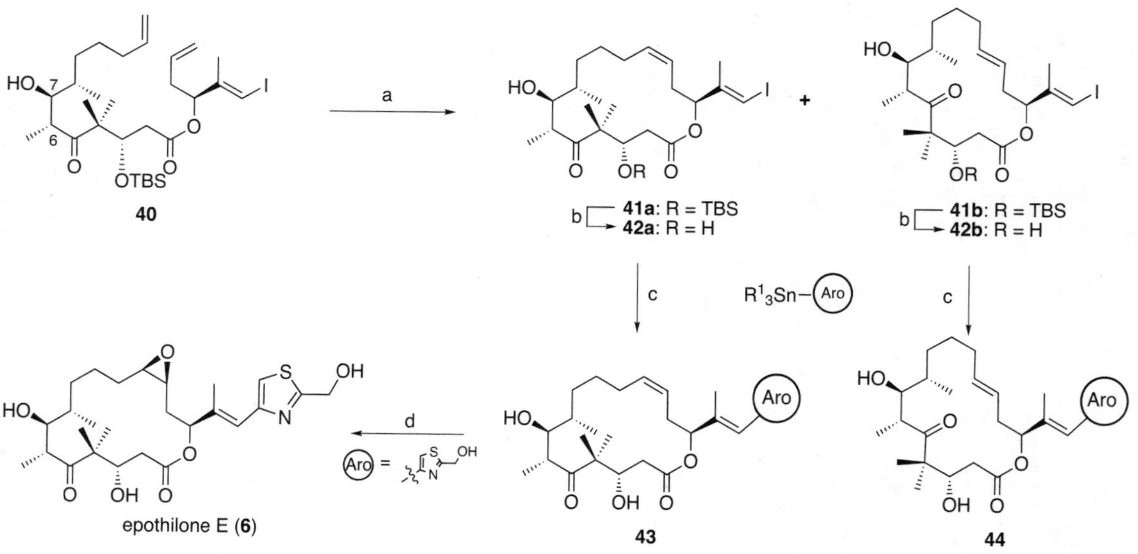

Scheme 9. Synthesis of epothilone E and side-chain modified epothilones via RCM and Stille coupling. (*a*) 0.1 equiv of RuCl₂(=CHPh)(PCy₃)₂ (3), CH₂Cl₂, 0.004 M, 25°C, 20 h, 65% (**41a:41b**=1.8:1); (*b*) HF•Py, THF, **43** (84%); **44** (85%); (*c*) R¹₃Sn(Aro), Pd(PPh₃)₄ or Pd(MeCN)₂Cl₂, 55–90%; (*d*) H₂O₂, CH₃CN, KHCO₃, 65%. Aro=aromatic (Nicolaou et al.) [20]

Table 1. Selected epothilone intermediates and analogs synthesized via ring-closing metathesis

RCM Precursor	RCM Product (yield, Z:E)[ref]	RCM Precursor	RCM Product (yield, Z:E)[ref]
29	**30** (86%, 1.7:1)[14b, 17] (86%, 1:2)[14b] (94%, 1:1)[16]	**57**	**58** (76%, 1:3)[21]
33a	**35** (86%, 1:3)[14b, 17]	**59**	**60** (72%, 4:3)[21]
33b	**36** (81%, 1:9)[14b, 17]	**61**	**62** (88%, 1:2.5)[21]
34	**27** (75%, 1:1)[18] (65%, 1:2)[14b, 17]	**63**	**64** (80%, 0:1)[13]
45	**46** (89%, 1:3.5)[13]	**65**	**66** (81%, 0:1)[13]
47	**48** (91%, 1:3)[18]	**67**	**68** (79%, 1:3)[18]
49	**50** (82%, 1:4)[18]	**69**	**70** (86%, 1:10)[18]
51	**52** (90%, 1:10)[18]	**71**	**72** (80%, 1:5)[14b, 17]
53	**54** (86%, 0:1)[13]	**73**	**74** (88%, 1:2)[14b, 17]
55	**56** (72%, 1:1)[21]	**75**	**76** (81%, 1:3)[26]

* All cyclisations were conducted using $RuCl_2(=CHPh)(PCy_3)_2$ (**3**) except for compound **30**, where $Mo(=CHCMe_2Ph)\{N(2,6-(i-Pr)_2C_6H_3)\}[OCMe(CF_3)_2]_2$ (**1**) was also used as the catalyst (result in second parentheses). R=TBS, R'=TPS, R"=TES.

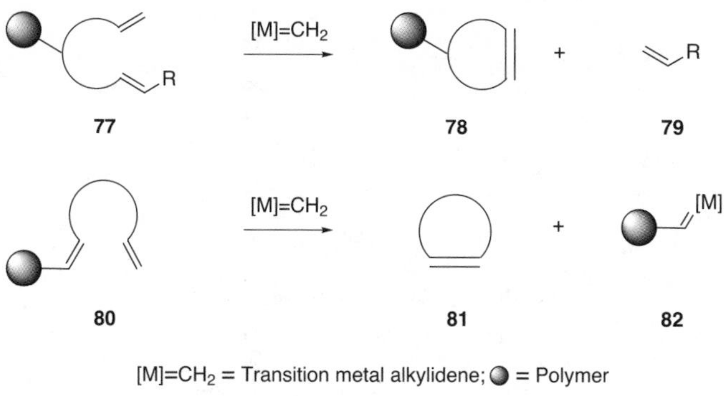

[M]=CH$_2$ = Transition metal alkylidene; ⬤ = Polymer

Scheme 10. Solid-phase synthesis using RCM

functionality will be cleaved from the solid support. Unwanted impurities should remain bound to the resin and therefore be separated readily by filtration. Unfortunately, this latter approach has a potential Achilles heel. Examination of the catalytic cycle (see Scheme 1) reveals that, following ring-closure, the active catalytic species will be a resin-bound carbene complex **82**. The presence of both reactant and catalyst on the polymer could potentially restrict reintroduction of the active species **82** into the catalytic cycle. A large amount of initiator may therefore be required to drive the reaction to completion. Despite these potential difficulties, a solid-phase synthesis of epothilone A (**4**) was recently reported by Nicolaou et al. [24] (Scheme 11).

The resin-bound trienes **83** (Scheme 11) were prepared in a similar fashion to the solution-phase studies (Sect. 2.2.2) and underwent tandem RCM resin-cleavage to liberate four macrolactones **84a,b** and **85a,b** in a combined yield of 52%. Although, as expected, a large amount of initiator **3** was required to effect this transformation, the procedure constituted a novel and efficient route to the epothilones which paved the way for the generation of a library of epothilone analogs. The library synthesis was achieved using the recently developed SMAR-T^a microreactors (SMART=single or multiple addressable radiofrequency tag) [25] (Scheme 12).

Each microreactor consists of a polymer-bound substrate and a radiofrequency encoded microchip enclosed within a small porous vessel. The radiofrequency tag allows the identity of the substrate contained within each microreactor to be established readily. Using this technology, the polymer-bound substrates **86** were individually elaborated, within separate microreactors, by sequential reactions with acids **87** and alcohols **88** in a similar way to the solution-phase processes [25c]. Each of the microreactors was then subjected to the tandem RCM resin-cleavage conditions employing initiator **3**. The products from each microreactor were obtained as a mixture of four compounds (**89–92**). The library of analogs prepared by this technique was then screened for biological activity [25c].

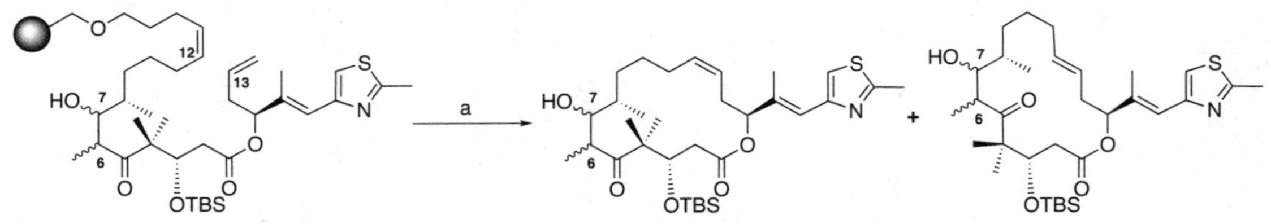

83

84a: 6R,7S
84b: 6S,7R

85a: 6R,7S
85b: 6S,7R

Scheme 11. Solid-phase synthesis of epothilones. (*a*) 0.75 equiv of $RuCl_2(=CHPh)(PCy_3)_2$ (**3**), CH_2Cl_2, 25°C, 48 h, 52% (**84a:84b:85a:85b**=3:3:1:3) (Nicolaou et al.) [24]

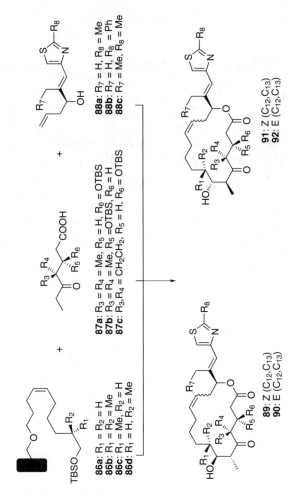

88a: R$_7$ = H, R$_8$ = Me
88b: R$_7$ = H, R$_8$ = Ph
88c: R$_7$ = Me, R$_8$ = Me

87a: R$_3$ = R$_4$ = Me, R$_5$ = H, R$_6$ = OTBS
87b: R$_3$ = R$_4$ = Me, R$_5$ =OTBS, R$_6$ = H
87c: R$_3$,R$_4$ = CH$_2$CH$_2$; R$_5$ = H, R$_6$ = OTBS

86a: R$_1$ = R$_2$ = H
86b: R$_1$ = R$_2$ = Me
86c: R$_1$ = Me, R$_2$ = H
86d: R$_1$ = H, R$_2$ = Me

91: Z (C$_{12}$,C$_{13}$)
92: E (C$_{12}$,C$_{13}$)

89: Z (C$_{12}$,C$_{13}$)
90: E (C$_{12}$,C$_{13}$)

Scheme 12. Generation of an epothilone library on solid-phase using SMART microreactors and the pool and split method (Nicolaou et al.) [25c]

2.4
Conclusion

The transformations discussed in Sects. 2.2–2.3 highlight several important features of the RCM process. Firstly, the compatibility of the ruthenium initiator **3** with a wide range of functional groups including epoxides, vinyl iodides, thiazoles and alcohols is demonstrated. The versatility of **3** is further illustrated in Sect. 2.3, where it is used to effect RCM of polymer-bound substrates. Previously, the molybdenum complex **1** has been reported to be more sensitive than **3** [19]. Experiments reported here are consistent with this view (Sect. 2.2.3) [14b].

The greater reactivity of terminal olefins compared to their more hindered di- and tri-substituted counterparts became evident in the model studies (Sect. 2.2.1) and in the total synthesis of epothilones A, B and E (Sects. 2.2.2–2.2.4). Suitably positioned disubstituted olefins can, however, participate in RCM reactions employing the molybdenum initiator **1** [19], and this is demonstrated in the total synthesis of epothilone B (**5**) (Sect. 2.2.3). As expected this transformation proved impossible using the ruthenium complex **3**.

The generality of the RCM approach for the synthesis of the C12,C13 bond of the epothilones is vividly illustrated (Sects. 2.2, 2.3, Table 1) and, in fact, no unsuccessful reactions using this disconnection strategy have been reported. Conversely, formation of the C9,C10 bond of a suitably functionalized epothilone intermediate has proved elusive despite extensive efforts [14]. Although the precise reason for this disparity remains to be elucidated, coordination of the incipient carbene complex with polar functionality proximal to the reacting olefin could be the factor inhibiting the latter approach [5]. This hypothesis has been proposed in similar unsuccessful attempts to achieve other large ring synthesis via RCM.

The preparation of the epothilone core using RCM is generally a high-yielding process. However, the Z:E-selectivity is mostly poor and, although a few trends have been observed (see Sect. 2.2.4), the ratio of products is largely unpredictable. The Z:E-selectivity is also dependent upon the solvent and initiator, although no conclusions concerning the nature of these subtle variances can be made at present.

Many of the observations discussed above clearly warrant further investigation. In particular methods for controlling the Z:E-selectivity of the ring-closure are of high priority. Recent reports [27] discussing the precise mode of action of the ruthenium initiators provide a solid foundation upon which these future studies can be built and success in this area will strengthen even further the already powerful RCM process.

3
Cyclopentadienyl Titanium Derivatives for Carbonyl Olefination/Olefin Metathesis

3.1
Introduction

Although the molybdenum and ruthenium complexes **1–3** have gained widespread popularity as initiators of RCM, the cyclopentadienyl titanium derivative **93** (Tebbe reagent) [28, 29] can also be used to promote olefin metathesis processes (Scheme 13) [28]. In a stoichiometric sense, **93** can be also used to promote the conversion of carbonyls into olefins [28b, 29]. Both transformations are thought to proceed via the reactive titanocene methylidene **94**, which is released from the Tebbe reagent **93** on treatment with base. Subsequent reaction of **94** with olefins produces metallacyclobutanes **95** and **97**. Isolation of these adducts, and extensive kinetic and labeling studies, have aided in the elucidation of the mechanism of metathesis processes [28].

3.2
Olefin Metathesis and Subsequent Intramolecular Carbonyl Olefination

Although use of the Tebbe reagent **93** in mechanistic studies (Sect. 3.1) has been extremely valuable, its synthetic utility in cross metathesis of 1,2-disubstituted olefins is limited. Unproductive cleavage of the initially formed metallacycles (**95,97**), back to the more stable alkylidene complex **94**, is the predominant process [28, 30b]. However, the relief of ring-strain associated with cleavage of a strained olefin can provide a driving force for productive metathesis. This concept has been exploited in the area of ring-opening metathesis polymerization (ROMP) and more recently in organic synthesis as illustrated by the elegant work of Grubbs [30] (Scheme 14). Ring-opening metathesis of the strained norbornene system **99** afforded the alkylidene complex **101** via metallacycle **100**. Intramolecular olefination of this reactive carbene intermediate with the ester functionality appended to the framework of the molecule, afforded tricyclic system **102**. Subsequent transformations led to the natural product, (±)-capnellene **104** [30] (Scheme 14).

Grubbs has reported a similar tandem olefin metathesis-carbonyl olefination process for the preparation of cyclic olefins [31]. In this case, treatment of a keto-olefin with the molybdenum alkylidene **1** at 20°C generates an intermediate alkylidene complex. Under these conditions, competing intermolecular olefination does not occur. However, intramolecular carbonyl olefination of the initially formed alkylidene complex can occur and this results in the formation of a cyclic olefin. This tandem sequence is illustrated by the transformation of keto-olefins **105** to cyclic ethers **106** (Scheme 15). Unfortunately, this protocol could not be applied to the formation of cyclic enol ethers **108** from esters **107** unless the more reactive (yet less readily prepared) tungsten initiator **109** was employed (Scheme 15) [31].

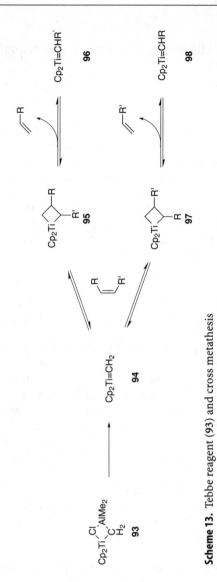

Scheme 13. Tebbe reagent (93) and cross metathesis

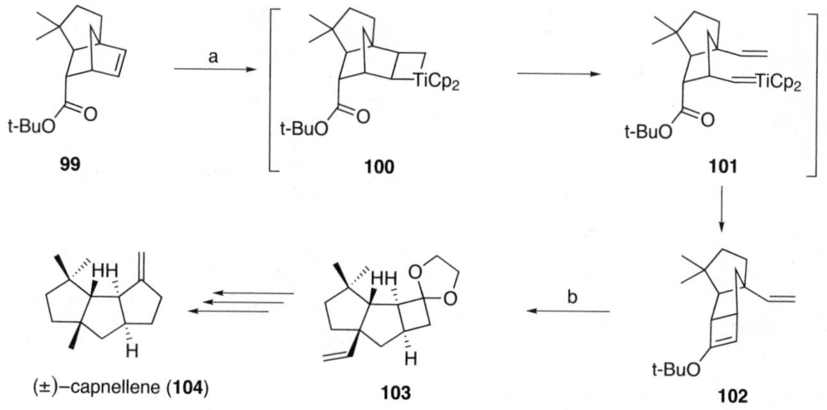

Scheme 14. Total synthesis of (±)-capnellene (**104**). (*a*) Tebbe reagent (**93**), DMAP, 25°C, then 90°C; (*b*) *p*-TsOH, (CH$_2$OH)$_2$, 81% overall (Grubbs et al.) [30]

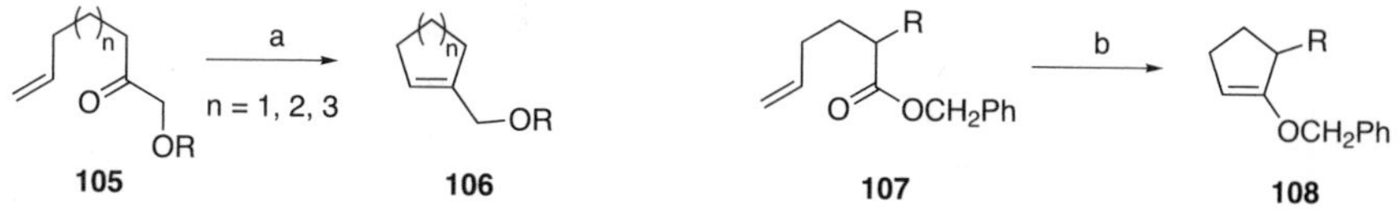

Scheme 15. Olefin metathesis-carbonyl olefination. (*a*) Mo(=CHCMe$_2$Ph){N[2,6-(*i*-Pr) $_2$C$_6$H$_3$]} [OCMe(CF$_3$)$_2$]$_2$ (**1**), 84–86%; (*b*) W(=CHCMe$_3$){N[2,6-(*i*-Pr)$_2$C$_6$H$_3$]}[OCMe(CF$_3$)$_2$] (**109**) (Grubbs et al.) [31]

3.3
Intermolecular Carbonyl Olefination and Subsequent Ring-Closing Metathesis

In addition to the intramolecular carbonyl olefination procedure outlined in Scheme 14, and unlike the molybdenum alkylidene **1** (Scheme 15), the Tebbe reagent **93** can promote intermolecular carbonyl methylenation at ambient temperature [28b, 29]. Recent studies by Petasis have demonstrated that dimethyl titanocene **110** (Cp$_2$TiMe$_2$) can also be used to effect this transformation [32]. The Petasis reagent **110**, which is less sensitive to air and moisture and therefore more easy to handle than its counterpart, generates the same reactive titanocene methylidene species **94** (cf. Scheme 13) and both reagents have been used to methylenate a variety of carbonyl compounds. In many cases, these reagents have proved superior to the more traditional Wittig-based methodologies. For example, the reaction between esters and the titanium species **93** and **110** generates enol ethers which are extremely useful intermediates for further synthetic transformations [32]. In this arena, Grubbs has recently demonstrated the prep-

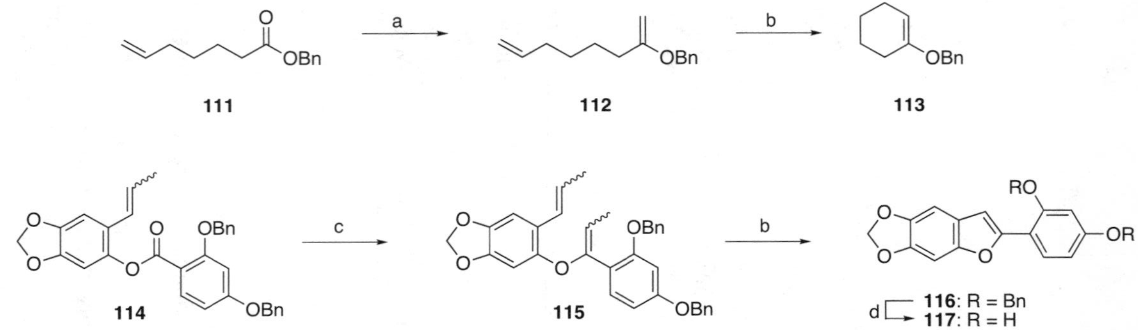

Scheme 16. Formation of enol ethers: synthesis of sophora compound I (117). (*a*) Tebbe reagent (**93**), toluene-THF, 60°C, 0.8 h, 81%; (*b*) 0.2 equiv of Mo(=CHCMe₂Ph){N[2,6-(*i*-Pr)₂C₆H₃]}[OCMe(CF₃)₂]₂ (**1**), 2–4 h, 84% (**113**), 85% (**116**); (*c*) MeCHBr₂, TiCl₄, Zn, TMEDA, PbCl₂, THF, 20°C, 11–20 h, 72%; (*d*) H₂, Pd/C, 96% (Grubbs et al.) [33]

aration of olefinic enol ethers **112** from their corresponding esters **111**. Subsequent reaction with molybdenum initiator **1** resulted in the formation of cyclic enol ethers **113** [33] (Scheme 16). This two-step carbonyl olefination-olefin metathesis procedure was applied to the preparation of the naturally occurring benzofuran Sophora compound I (**117**) via diene **115**, although in this example the initial olefination was achieved using the Utimoto-Takai protocol [33].

Recently, Nicolaou and coworkers have devised a novel, one-pot strategy for the direct transformation of acyclic olefinic esters to cyclic enol ethers [34]. Unlike the molybdenum alkylidene **1** (see Sect. 3.2), initial reaction between the Tebbe reagent **93** and an olefinic ester results in rapid carbonyl olefination to afford a diene intermediate. Subsequent heating initiates RCM to afford the desired cyclic product (Scheme 17).

The impetus for developing this technology was provided by the continuing need for methodology amenable to the preparation of polycyclic arrays that constitute the frameworks of numerous polyether natural products such as the brevetoxins, ciguatoxin and maitotoxin **124** (Fig. 3).

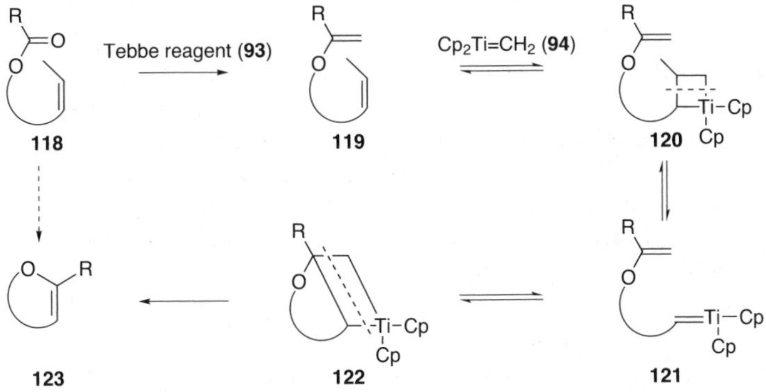

Scheme 17. Titanium-mediated metathesis strategy for the conversion of olefinic esters (**118**) to cyclic enol ethers (**123**) (Nicolaou et al.) [34]

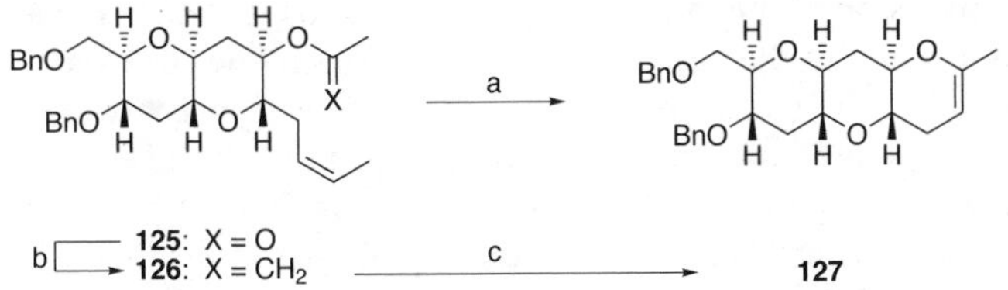

Scheme 18. The conversion of olefinic ester **125** to cyclic enol ether **127**. (*a*) 4.0 equiv of Tebbe reagent (**93**), 25°C, 20 min; then reflux, 5 h, 71%; (*b*) 1.3 equiv of Tebbe reagent (**93**), 25°C, 20 min, 77%; (*c*) 2.0 equiv of Tebbe reagent (**93**), 25°C, 20 min; then reflux, 3 h, 65% (Nicolaou et al.) [34a]

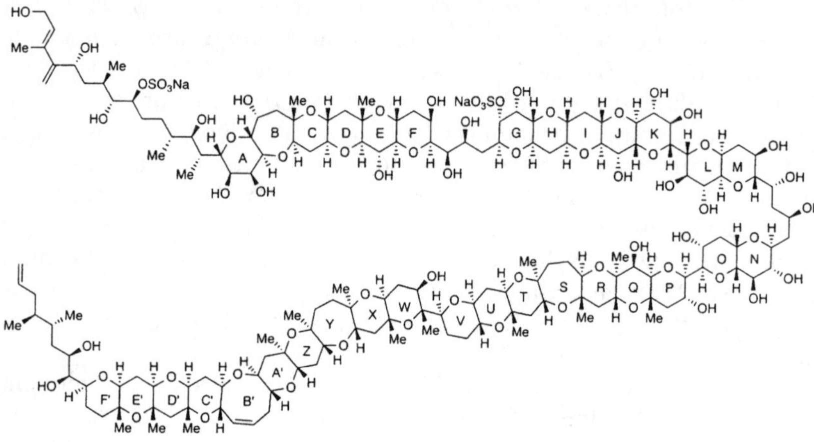

Fig. 3. Structure of maitotoxin (124)

Preliminary investigations in this area involved treatment of olefinic ester **125** with a large excess (4 equiv) of the Tebbe reagent **93** (Scheme 18) [34a]. After 20 min at 25°C, the mixture was heated at reflux for 5 h. This resulted in the formation of tricyclic enol ether **127** in 71% overall yield. If only 1.3 equiv of Tebbe reagent **93** was employed and the reaction stopped after 20 min at 25°C, the olefinic enol ether **126** could be isolated in 77% yield. The proposed intermediacy of diene **126** in the initial tandem sequence was validated by its subsequent conversion into the cyclic enol ether **127** under the original reaction conditions [34a].

These initial forays verified the utility of the process and provided mechanistic information which set the stage for further exploration into the scope of the process. Successful reactions with a variety of substrates demonstrated the generality of the process, with 6- and 7-membered cyclic enol ethers being most accessible (Scheme 19). Preservation of the trisubstituted olefin present in the complex substrate **132** indicates the chemoselective nature of the process [34a].

The efficacy of the procedure for the synthesis of highly complex systems, and the utility of the enol ether functionality obtained in the reaction is demonstrated by the conversion of the intricate polyether **134** to enol ether **135** and its subsequent elaboration to the hexacyclic system **138** (Scheme 20) [34a].

A limitation to the use of the Tebbe reagent **93** was observed during the attempted conversion of substrates **139** and **142** to the tricyclic systems **141** and **144** respectively (Scheme 21). The major products from these reactions were olefinic alcohols **140** and **143**. These products presumably resulted from sequential hydrolysis and olefination of the initially formed cyclic enol ethers. The problem associated with these sensitive substrates was overcome through use of the less Lewis-acidic Petasis reagent **110,** which provided access to the desired products **141** and **144** [34a].

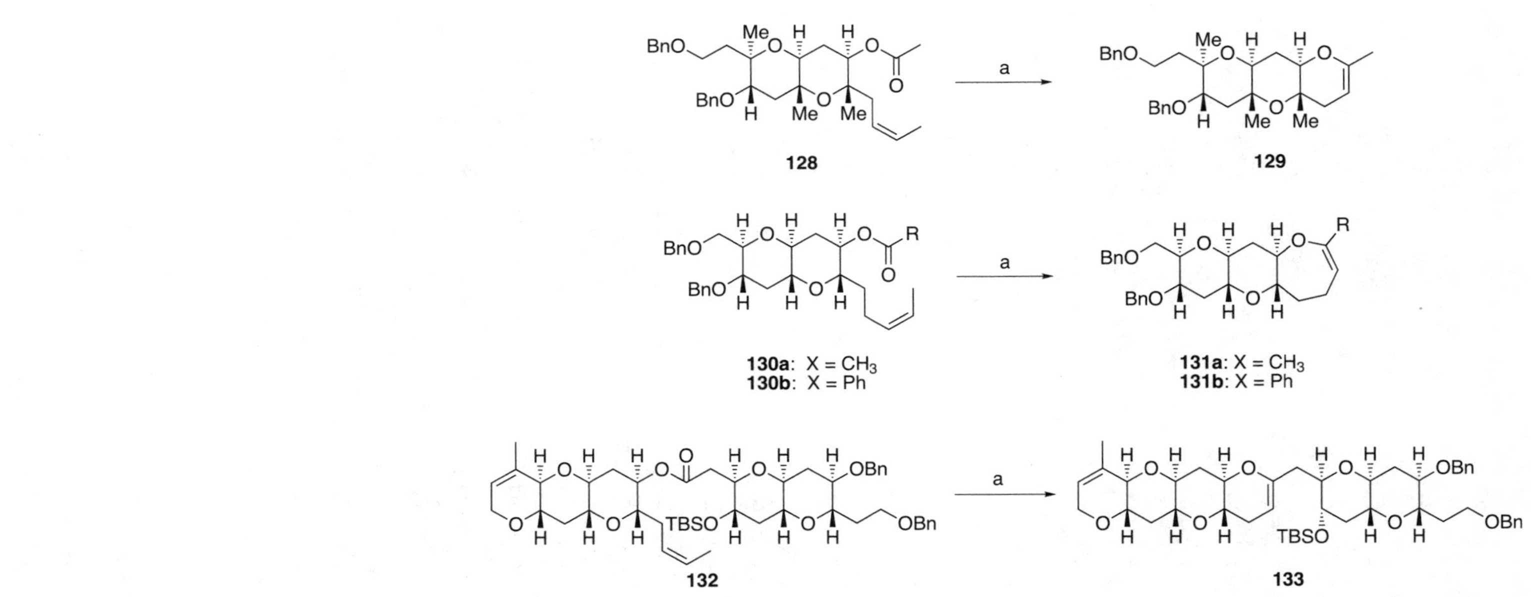

Scheme 19. Titanium-mediated metathesis strategy for the conversion of olefinic esters to 6- and 7-membered cyclic enol ethers. (*a*) 4.0 equiv of Tebbe reagent (**93**), THF, 25°C, 20 min; then reflux, 2–8 h, 64% (**129**), 45% (**131a**), 32% (**131b**), 45% (**133**) (Nicolaou et al.) [34a]

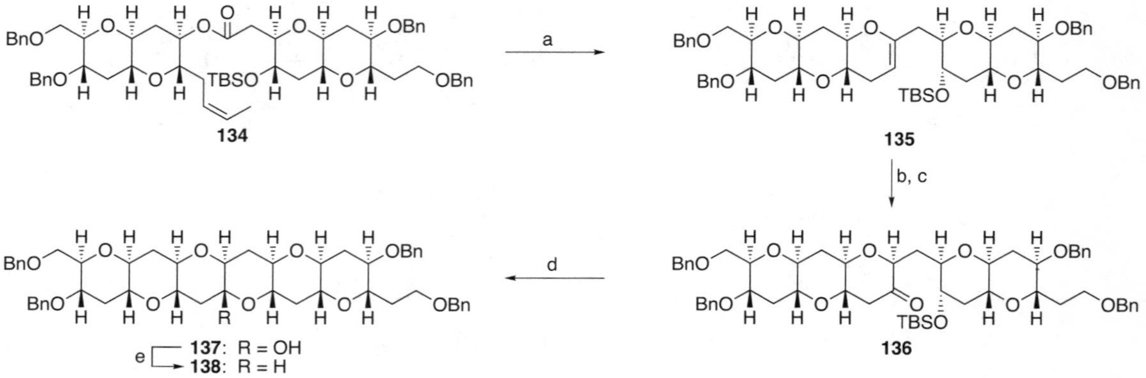

Scheme 20. Synthesis of hexacyclic polyether **138**. (*a*) 4.0 equiv of Tebbe reagent (**93**), THF, 25°C, 0.5 h; then reflux, 10 h, 71%; (*b*) BH₃; then H₂O₂; (*c*) Dess-Martin reagent, 60% for two steps; (*d*) TBAF, 60%; (*e*) Et₃SiH, BF₃,Et₂O, 91% (Nicolaou et al.) [34a]

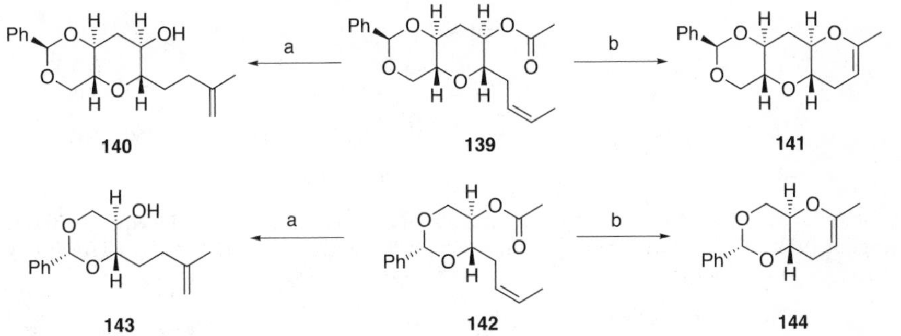

Scheme 21. Titanium-mediated metathesis strategy for the synthesis of cyclic enol ethers and hydroxy olefins. (*a*) Tebbe reagent (**93**), THF, reflux, 41% (**140**), 41% (**143**); (*b*) Petasis reagent (**110**), THF, reflux, 60% (**141**), 30% (**144**) (Nicolaou et al.) [34a]

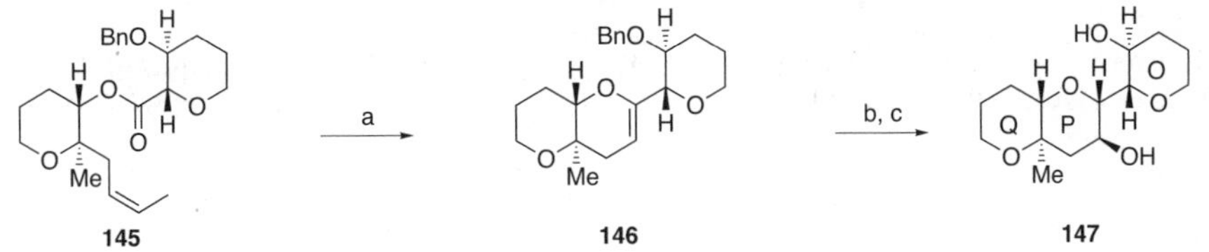

Scheme 22. Synthesis of OPQ ring system (**146**) of maitotoxin (**124**). (*a*) 4.0 equiv of Tebbe reagent (**93**), THF, 25°C, 0.5 h; then reflux, 4 h, 54%; (*b*) BH$_3$; then H$_2$O$_2$, 89%; (*c*) H$_2$, Pd/C, 91% (Nicolaou et al.) [34b]

Examination of the impressive and imposing molecular structure of maitotoxin **124** (Fig. 3) reveals three regions in which carbon–carbon bonds link the more common polyether domains. The preparation of model systems for these unusual sectors of the molecule was achieved using the novel one-pot RCM methodology described above [34b]. For example, preparation of the OPQ fragment **147** was achieved from the olefinic ester **145**, which was readily accessible through conventional procedures (Scheme 22). Smooth transformation into cyclic enol ether **146** was achieved using the Tebbe reagent **93**. Subsequent elaboration of **146** was accomplished through efficient and stereoselective oxidation, followed by removal of the benzyl protecting group, affording **147** [34b].

Preparation of the bridging fragment **150** (Scheme 23) followed a similar pathway. In this case, stereoselective reduction of the cyclic enol ether **149** formed by the RCM of **148** was achieved using Et$_3$SiH and TFA, leading after desilylation to the WVU ring system **150** [34b].

Finally, production of tricyclic array **153** (Scheme 24) required cyclization of bicyclic ester **151** in which additional oxygenation was present in the olefinic appendage. The successful conversion to enol ether **152** demonstrated further the power of the method and led to the JKL fragment **153** [34b].

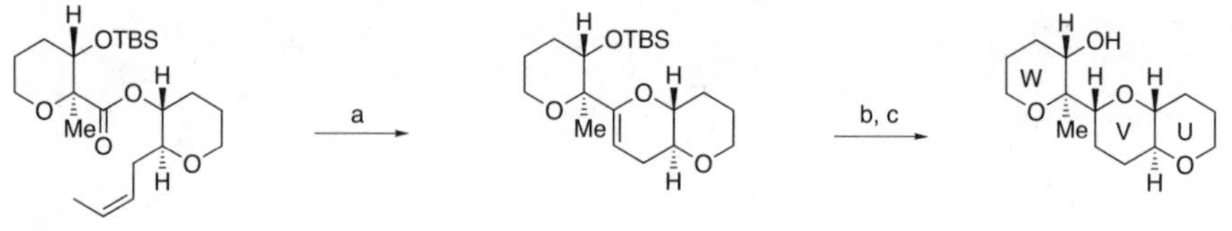

Scheme 23. Synthesis of UVW ring system (150) of maitotoxin (124). (*a*) 4.0 equiv of Tebbe reagent (93), THF, 25 °C, 0.5 h; then reflux, 4 h, 36%; (*b*) TFA, Et$_3$SiH, 80%; (*c*) TBAF, 91% (Nicolaou et al.) [34b]

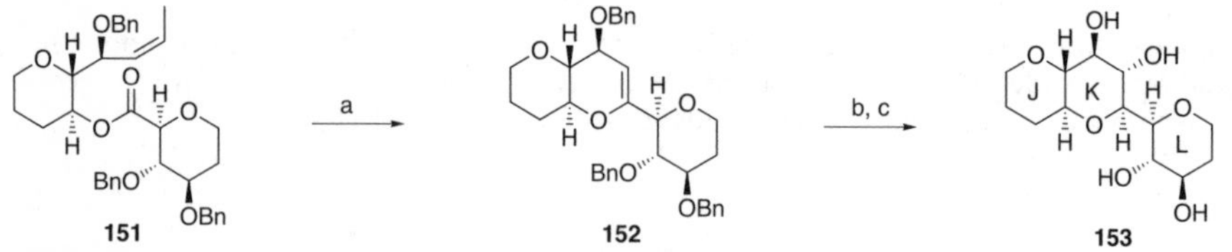

Scheme 24. Synthesis of JKL ring system (147) of maitotoxin (124). (*a*) 10.0 equiv of Petasis reagent (110), THF, reflux, 12 h, 20%; (*b*) BH$_3$; then H$_2$O$_2$, 77%; (*c*) H$_2$, Pd/C, 98% (Nicolaou et al.) [34b]

3.4
Conclusion

Cyclopentadienyl titanium complexes, in particular the Tebbe reagent 93, have played an important role in both olefin metathesis and carbonyl olefination processes. Recently, several examples illustrating the consecutive use of these transformations to generate cyclic systems have been reported. In the first example, Scheme 14, the Tebbe reagent 93 was used to generate an intermediate alkylidene via ring-opening metathesis of a strained olefin. This one-pot process is concluded by intramolecular carbonyl olefination to afford a cyclic enol ether. Unfortunately, this methodology is of limited applicability since a strained olefin is required to promote the initial ring-opening step. However, the elegant synthesis of (±)-capnellene (104) described by Grubbs (Scheme 14) illustrates the power of the method in suitable systems. In a similar approach, treatment of a keto-olefin with the molybdenum alkylidene 1 generated an intermediate alkylidene which could participate in intramolecular carbonyl olefination to afford cyclic olefins (Scheme 15). However, application of this approach to the preparation of cyclic enol ethers is of limited synthetic value since a less practical initiator is required.

An alternative approach involves a two-step procedure, in which carbonyl olefination, using the Tebbe reagent 93, generates an acyclic enol ether-olefin (Scheme 16). In this case, subsequent RCM using molybdenum alkylidene 1 proceeds to give cyclic enol ethers. An efficient, one-pot carbonyl olefination-RCM approach has been developed by Nicolaou et al. for the formation of cyclic enol ethers directly from acyclic olefinic esters (Scheme 17). The strategy, which uses either the Tebbe or Petasis reagents, has been used in the preparation of highly complex polyether ring systems. The potential of this methodology to complex molecule construction is obvious and will no doubt be realized in the near future.

Acknowledgements

We wish to thank our collaborators whose names appear in the references. We gratefully acknowledge the National Institutes of Health USA, Merck Sharp & Dohme, Schering Plough, Pfizer, Hoffmann La Roche, Glaxo, Rhone-Poulenc Rorer, Amgen, Novartis, CaP CURE, the Skaggs Institute for Chemical Biology, and the George Hewitt Foundation for financial support.

4
References

1. (a) Schuster M, Blechert S (1997) Angew Chem Int Ed Eng 36:2036. (b) Grubbs RH, Miller SJ, Fu GC (1995) Acc Chem Res 28:446. (c) Schmalz H-G (1995) Angew Chem Int Ed Eng 34:1833
2. (a) Villemin D (1980) Tetrahedron Lett 21:1715. (b) Tsuji J, Hashiguchi S (1980) Tetrahedron Lett 21:2955
3. Schrock RR, Murdzek JS, Bazan GC, Robbins J, DiMare M, O'Regan M (1990) J Am Chem Soc 112:3875
4. (a) Nguyen ST, Johnson LK, Grubbs, RH (1992) J Am Chem Soc 114:3974. (b) Schwab P, France MB, Ziller JW, Grubbs RH (1995) Angew Chem Int Ed Engl 34:2039
5. Fürstner A, Langemann K (1997) Synthesis 792
6. (a) Höfle G, Bedorf N, Steinmetz H, Schomburg D, Gerth K, Reichenbach H (1996) Angew Chem Int Ed Engl 35:1567. (b) Höfle G, personal communication
7. Nicolaou KC, Dai W-M, Guy RK (1994) Angew Chem Int Ed Engl 33:15
8. (a) Bollag DM, McQueney PA, Zhu J, Hensens O, Koupal L, Liesch J, Goetz M, Lazarides E, Woods CM (1995) Cancer Res 55:2325. (b) Kowalski RJ, Giannakakou P, Hamel E (1997) J Biol Chem 272:2534. (c) Gerth K, Bedorf N, Höfle G, Irschik H, Reichenbach, H (1996) J Antibiot 49:560
9. Borer BC, Deerenberg S, Bieräugel H, Pandit UK (1994) Tetrahedron Lett 35:3191
10. (a) Xu Z, Johannes CW, Houri AF, La DS, Cogan DA, Hofilena GE, Hoveyda AH (1997) J Am Chem Soc 119:10302. (b) Xu Z, Johannes CW, Salman SS, Hoveyda AH (1996) J Am Chem Soc 118:10926. (c) Houri AF, Xu Z, Cogan DA, Hoveyda AH (1995) J Am Chem Soc 117:2943
11. (a) Fürstner A, Langemann K (1997) J Am Chem Soc 119:9130. (b) Fürstner A, Müller T (1997) Synlett 1010. (c) Fürstner A, Langemann K (1996) J Org Chem 61:3942. (d) Fürstner A, Langemann K (1996) J Org Chem 61:8746. (e) Fürstner A, Kindler N (1996) Tetrahedron Lett 37:7005
12. Nicolaou KC, He Y, Vourloumis D, Vallberg H, Yang Z (1996) Angew Chem Int Ed Engl 35:2399
13. Nicolaou KC, He Y, Vourloumis D, Vallberg H, Roschangar F, Sarabia F, Ninkovic S, Yang Z, Trujillo JI (1997) J Am Chem Soc 119:7960
14. (a) Bertinato P, Sorensen EJ, Meng D, Danishefsky SJ (1996) J Org Chem 61:8000. (b) Meng D, Bertinato P, Balog A, Su D-S, Kamenecka T, Sorensen EJ, Danishefsky SJ (1997) J Am Chem Soc 119:10073
15. Yang Z, He Y, Vourloumis D, Vallberg H, Nicolaou KC (1997) Angew Chem Int Ed Engl 36:166

16. Schinzer D, Limberg A, Bauer A, Böhm OM, Cordes M (1997) Angew Chem Int Ed Engl 36:523
17. (a) Balog A, Meng D, Kamenecka T, Bertinato P, Su D-S, Sorensen EJ, Danishefsky SJ (1996) Angew Chem Int Ed Engl 35:2801. (b) Meng D, Su D-S, Balog A, Bertinato P, Sorensen EJ, Danishefsky SJ, Zheng Y-H, Chou FC, He L, Horwitz SB (1997) J Am Chem Soc 119:2733
18. Nicolaou KC (unpublished results)
19. Kirkland TA, Grubbs RH (1997) J Org Chem 62:7310
20. (a) Nicolaou KC, He Y, Roschangar F, King NP, Vourloumis D, Li T (1998) Angew Chem Int Ed Engl 37:84
    (b) Nicolaou KC, King NP, Finlay MRV, He Y, Roschangar F, Vourlomis D, Vallberg H, Sarabia F, Ninkovic S, Hepworth D (1998) Biorgan Med Chem (in press).
21. (a) Nicolaou KC, Vallberg H, King NP, Roschangar F, He Y, Vourloumis D, Nicolaou CG (1997) Chem Eur J 3:1957
22. (a) Balkenhohl F, von dem Bussche-Hünnefeld C, Lansky A, Zechel C (1996) Angew Chem Int Ed Engl 35:2288 and references cited therein. (b) Hermkens PHH, Ottenheijm HCJ, Rees DC (1997) Tetrahedron 53:5643
23. (a) Piscopio AD, Miller JF, Koch K (1997) Tetrahedron Lett 38:7143. (b) Schuster M, Pernerstorfer J, Blechert S (1996) Angew Chem Int Ed Eng 35:1979. (c) Peters J-U, Blechert S (1997) Synlett 348. (d) Pernerstorfer J, Schuster M, Blechert S (1997) J Chem Soc Chem Commun 1949. (e) van Maarseveen JH, den Hartog, JAJ, Engelen V, Finner E, Visser G, Kruse CG (1996) Tetrahedron Lett 37:8249. (f) Miller SJ, Blackwell HE, Grubbs RH (1996) J Am Chem Soc 118:9606
24. Nicolaou KC, Winssinger N, Pastor J, Ninkovic S, Sarabia F, He Y, Vourloumis D, Yang Z, Li T, Giannakakou P, Hamel E (1997) Nature 387:268
25. (a) Nicolaou KC, Xiao X-Y, Parandoosh Z, Senyei A, Nova MP (1995) Angew Chem Int Ed Engl 34:2289. (b) Czarnik AW, Nova MP (1997) Chem Br 33(10):39. (c) Nicolaou KC, Vourloumis D, Li T, Pastor J, Winssinger N, He Y, Ninkovic S, Sarabia F, Vallberg H, Roschangar F, King NP, Finlay MRV, Giannakakou P, Verdier-Pinard P, Hamel E (1997) Angew Chem Int Ed Engl 36:2097
26. Taylor RE, Haley JD (1997) Tetrahedron Lett 38:2061
27. (a) Dias EL, Nguyen ST, Grubbs RH (1997) J Am Chem Soc 119:3887. (b) Tallarico JA, Bonitatebus Jr PJ, Snapper ML (1997) J Am Chem Soc 119:7157
28. (a) Tebbe FN, Parshall GW, Ovenall DW (1979) J Am Chem Soc 101:5074. (b) Brown-Wensley KA, Buchwald SL, Cannizzo L, Clawson L, Ho S, Meinhardt D, Stille JR, Straus D, Grubbs RH (1983) Pure Appl Chem 55:1733. (c) Ott KC, Grubbs RH (1981) J Am Chem Soc 103:5922. (d) Lee JB, Ott KC, Grubbs RH (1982) J Am Chem Soc 104:7491
29. (a) Tebbe FN, Parshall GW, Reddy GS (1978) J Am Chem Soc 100:3611. (b) Pine SH, Pettie, RJ, Geib GD, Cruz SG, Gallego CH, Tijerina T, Pine RD (1985) J Org Chem 50:1212. (c) Pine SH, Zahler R, Evans DA, Grubbs RH (1980) J Am Chem Soc 102:3271
30. (a) Stille JR, Santarsiero BD, Grubbs RH (1990) J Org Chem 55:843. (b) Stille JR, Grubbs RH (1986) J Am Chem Soc 108:855
31. Fu GC, Grubbs RH (1993) J Am Chem Soc 115:3800
32. (a) Petasis NA, Lu S-P, Bzowej EI, Fu D-K, Straszewski JP, Akritopoulou-Zanze I, Patane MA, Hu Y-H (1996) Pure Appl Chem 68:667. (b) Petasis NA, Hu Y-H (1997) Curr Org Chem 1:249
33. Fujimura O, Fu GC, Grubbs RH (1994) J Org Chem 59:4029
34. (a) Nicolaou KC, Postema MHD, Claiborne CF (1996) J Am Chem Soc 118:1565. (b) Nicolaou KC, Postema MHD, Yue EW, Nadin A (1996) J Am Chem Soc 118:10335

Catalytic Ring-Closing Metathesis and the Development of Enantioselective Processes

Amir H. Hoveyda

Catalytic ring-closing metathesis makes available a wide range of cyclic alkenes, thus rendering a number of stereoselective olefin functionalizations practical. The availability of effective metathesis catalysts has also spawned the development of a variety of methods that prepare specially-outfitted diene substrates that can undergo catalytic ring closure. The new metathesis catalysts have already played a pivotal role in a number of enantioselective total syntheses.

Keywords: Catalytic metathesis, Enantioselective synthesis, Catalytic kinetic resolution, Catalytic carbomagnesation, Catalytic rearrangement, 2-Substituted chromenes, Catalytic macrocyclization

1
Introduction

Investigations in a number of research groups during the past several years have led to the development of catalytic ring-closing metathesis (RCM) as a reliable and practical method for selective carbon–carbon double bond formation [1]. The rise of catalytic RCM as a prominent transformation in synthetic organic chemistry is largely due to the advent of the Ru-based catalysts of Grubbs (1) [2] and the Mo-based complex (2), first introduced and developed by Schrock [3] (Scheme 1). These catalysts are relatively straightforward to prepare in large quantities, and are tolerant of a number of functional groups. Together, these two remarkable transition metal complexes give access to an impressive range of unsaturated carbo- and heterocycles.

It is therefore not a surprise that the availability of catalysts such as 1 and 2 has spawned a thriving and important area of research. The success of technologies that utilize catalytic RCM is based on three additional factors: (1) Catalytic RCM can be carried out efficiently on readily accessible diene precursors, thus enabling the synthetic chemist to gain easy access to a gamut of the requisite unsaturated starting materials, suitable for further functionalization. (2) Ring-closure is often stereoselective. As will be detailed below, this issue is particularly critical with the Mo-catalyzed preparation of trisubstituted olefins; this is in contrast to disubstituted olefin mixtures, where simple hydrogenation removes any complications arising from formation of stereoisomeric (olefin) mixtures. (3) Since small amounts of the RCM catalysts are often sufficient (high turnovers) and the cyclic products are formed cleanly, with volatile alkenes as the only side products (e.g., ethylene), can subsequent functionalizations be carried out in the same vessel. This tandem strategy and, in many cases, fully-catalytic process leads to minimization of waste solvents, and obviates the requirement for an interim purification procedure, allowing the synthesis to be carried out more expeditiously and less expensively.

In 1993, we began to incorporate catalytic RCM into a select number of our programs in reaction development and enantioselective organic synthesis. In

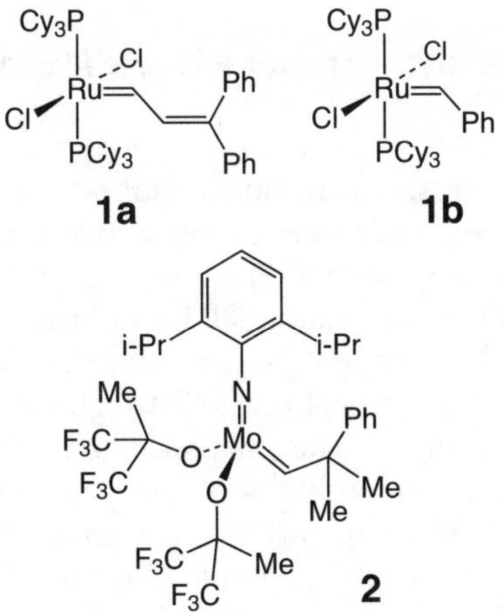

Scheme 1. Metathesis catalysts introduced by Grubbs (**1a–b**) and Schrock (**2**)

retrospect, the involvement of RCM in our research objectives can be divided into two categories:

1. Products of catalytic RCM were formerly not attainable by any other means, or in such a selective and efficient manner. With these unsaturated substrates easily accessible, myriad reaction technologies have emerged that involve the regio- and stereoselective functionalization of such adducts. In the absence of catalytic RCM, these processes would either remain undeveloped or would be, for the most part, categorized as "interesting but unpractical" methods in synthesis.

2. Another group of reactions that emerged as supporting companions to catalytic RCM reactions are transformations that provide specially outfitted and useful diene substrates for the metathesis process. Such compounds, in the presence of **1a**, **1b** or **2**, can be converted to otherwise difficult-to-make organic molecules. Research activity in this area has led to the design of synthesis methods that elevate the utility of **1** and **2** beyond ring closure.

This review article offers an overview of stereo- and regioselective processes, developed in our laboratories, that are carried out in conjunction with various catalytic RCM reactions. As mentioned above, several reaction technologies involve stereoselective functionalization of the RCM products, and others were developed such that subsequent reactions effected by a metathesis catalyst promote the formation of the desired target molecules selectively and efficiently.

2
Catalytic RCM and Zr-Catalyzed Enantioselective Alkylation of Unsaturated Heterocycles

During the past several years, research in our laboratories has focused on the development of efficient catalytic and enantioselective addition of alkylmetals to alkenes. One area of investigation that has been fruitful in this regard is the Zr-catalyzed addition of alkylmagnesium halides to unsaturated heterocycles [4]. As discussed below, the Zr-catalyzed process allows one to obtain a range of organic molecules in excellent optical purity through catalytic asymmetric synthesis and kinetic resolution [5]. The emergence of catalytic RCM proved to be crucial to the development and utility of these transition-metal catalyzed reactions, as it offered a reliable and convenient method for the preparation of a much larger number of the requisite olefinic substrates.

2.1
Enantioselective Synthesis by Tandem Catalytic RCM and Catalytic Alkylation

In 1993, we reported that various unsaturated heterocycles can be alkylated with Et-, nPr- and nBuMgCl in the presence of optically pure (EBTHI)ZrCl$_2$ (**3a**) or (EBTHI)Zr-binol (**3b**) to afford the derived unsaturated products in >90% ee (cf. **5→6**, Scheme 2) [4a]. Many of the simpler five- and six-membered starting materials are available commercially or can be prepared by established procedures. In contrast, catalytic enantioselective reactions involving unsaturated medium ring heterocycles were not a trivial undertaking; the synthesis of these olefinic substrates, by the extant methods, was prohibitively cumbersome.

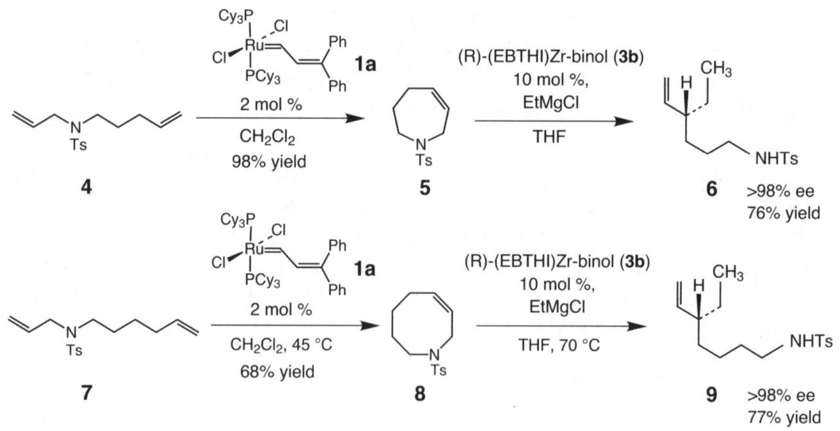

Scheme 2. Ru-catalyzed RCM efficiently provides substrates required for the Zr-catalyzed enantioselective alkylation

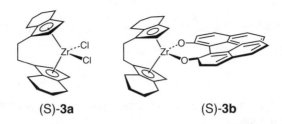

(S)-**3a**   (S)-**3b**

As the examples in Scheme 2 illustrate, the emergence of catalytic RCM proved instrumental in rendering the Zr-catalyzed C-C bond-forming reaction a more viable method in asymmetric synthesis [5d]. Catalytic RCM of dienes **4** and **7**, effected by 2 mol% of Ru catalyst **1a**, leads to the formation of **5** and **8** in high yield. Subsequent Zr-catalyzed alkylation of the resulting heterocycles in the presence of 10 mol% **3b** delivers unsaturated amides **6** and **9** in the optically pure form (>98% ee) and in 76% and 77% isolated yield, respectively (Scheme 2).

It is important to note that the Ru-catalyzed RCM and the Zr-catalyzed resolution can be carried out in a single vessel, without recourse to intermediate isolation. The unsaturated medium-ring amides **5** and **8** can be subjected to 10 mol% of the chiral Zr catalyst and EtMgCl, in the same flask, to afford unsaturated **6** and **9** in 81% and 54% isolated yield, respectively. As depicted in Eq. 1, a similar tandem diene metathesis/ethylmagnesation can be carried out on ether **10**, leading to the formation of unsaturated chiral alcohol **11** in 73% yield and >99% ee.

$$\text{2 mol \% } \mathbf{1b}, \text{ THF};$$
$$\xrightarrow{\hspace{1cm}}$$
$$\text{EtMgCl,}$$
$$\text{10 mol \% (R)-}\mathbf{3a}$$

10   73% yield   **11** >99% ee   (1)

The successful catalytic RCM involving diene **7** (Scheme 2) is noteworthy [5d], since, unlike the syntheses of seven-membered ring systems (e.g., reactions with **4** and **10**), construction of the derived eight-membered heterocycle is expected to proceed inefficiently [6]. Furthermore, this result merits attention in light of recent reports that in catalytic metathesis reactions that lead to eight-membered rings, the presence of rigid frameworks can play a critical role in the efficiency of ring formation [6a]. As expected, and in contrast to the unusually facile ring closure of amide derivative **7** to afford **8** (Scheme 2), treatment of **12** with 2 mol% **1a** in refluxing CH_2Cl_2 (Eq 2) leads to the formation of a variety of byproducts (^{1}H NMR analysis; reaction is even less efficient in THF). It is plausible that in the Ru-catalyzed synthesis of **8**, the sterically demanding Ts group, has a favorable conformational effect [7] on the diene substrate, leading to a significantly more facile ring closure [8].

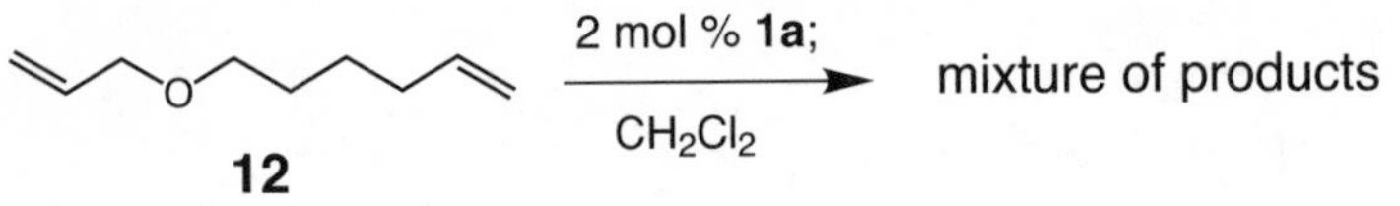

$$2 \text{ mol } \% \text{ 1a;} \xrightarrow{\text{CH}_2\text{Cl}_2} \text{mixture of products} \quad (2)$$

12

2.2
Enantioselective Synthesis of Unsaturated Heterocycles by Tandem Catalytic RCM-Catalytic Kinetic Resolution

Along with the Zr-catalyzed asymmetric alkylation of heterocycles, we also established in 1994 that the catalytic ethylmagnesation can be used to effect the kinetic resolution of a range of chiral unsaturated pyrans, oxepins and oxocenes [5a]. Although a select number of the racemic dihydropyrans were readily available by the existing methods, the seven- and eight-membered ring analogs were not. It was thus crucial that catalytic RCM allowed easy access to a significantly wider range of substrates that could then be efficiently resolved by the Zr-catalyzed protocol. The two examples shown in Scheme 3 are illustrative [5d]. In both instances, the racemic substrates (**14** and **17**) are prepared from simple dienes; subsequent kinetic resolution delivers these medium ring heterocycles in the non-racemic form with high optical purity. It is also worthy of note that, in addition to the recovered starting material, the alkylation products (e.g., (S)-**15** in Scheme 3) can be isolated with excellent enantioselectivity.

The catalytic RCM and kinetic resolution can be carried out in a single vessel as well. This is particularly important for the practical utility of the Zr-catalyzed resolution: Because the best theoretical yield in a classical resolution is 50%, it is imperative that the racemic substrate is prepared readily (or 50% material loss will be too costly). In this instance, the racemic substrate is not only obtained efficiently, it is synthesized in a catalytic manner and need not even be isolated prior to the resolution. Two representative examples are illustrated in Scheme 4 [5a]. The tandem catalytic RCM, leading to *rac*-**19** and its subsequent catalytic resolution proceeds with excellent efficiency: the one-vessel, two-stage process

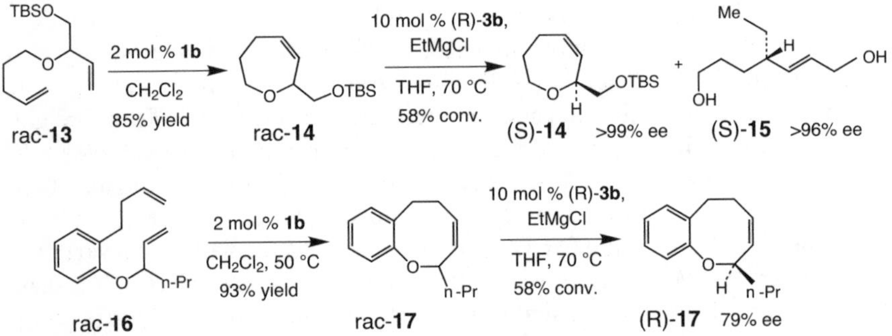

Scheme 3. Ru-catalyzed RCM efficiently provides substrates required for the Zr-catalyzed kinetic resolution

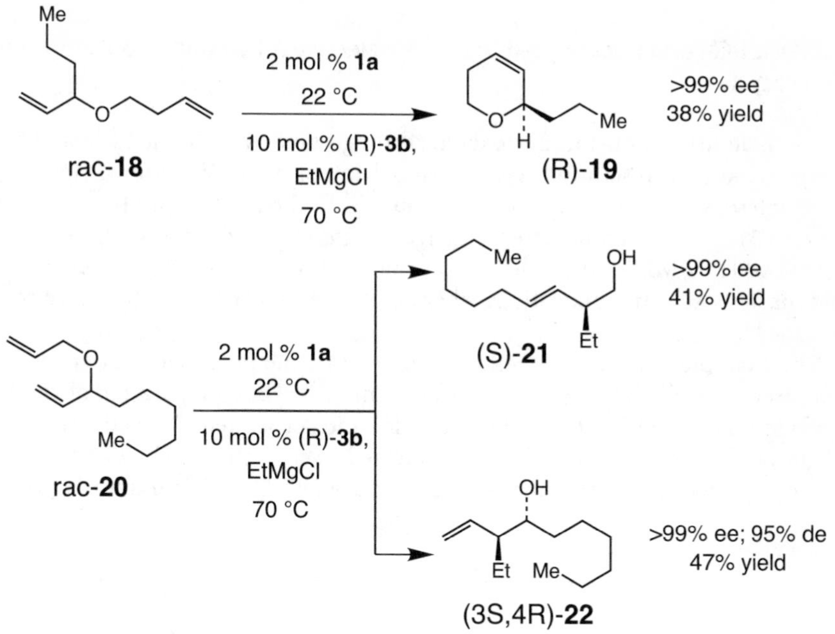

Scheme 4. Tandem Ru-catalyzed RCM and Zr-catalyzed kinetic resolution can be carried in a single vessel to afford a variety of optically pure materials

requires a total of 2–3 h of reaction time and the overall yield of the optically pure (R)-19 after silica gel chromatography from diene rac-18 is 38% (maximum yield is 50%).

As also shown in Scheme 4, when rac-20, prepared from simple alkylation of the corresponding allylic alcohol with allylbromide, is treated with 2 mol% 1a and the reaction mixture is then treated with 10 mol% (R)-3b and five equivalents of EtMgCl at 70°C, (S)-21 and (3S,4R)-22 are obtained in >99% ee and 41% and 47% yield after silica gel chromatography, respectively (chiral GLC analysis) [5b]. Thus, from simple starting materials and in a single vessel, compounds of excellent optical purity can be obtained efficiently (88% yield). As before, the racemic dihydrofuran intermediate need not be isolated.

The resolution of rac-20 represents a less common form of catalytic kinetic resolution (parallel kinetic resolution) [9]. In conventional kinetic resolution, one substrate enantiomer reacts preferably to leave behind the unreacted isomer in high optical purity (e.g., rac-18→(R)-19 in Scheme 4). In this instance, *both* starting material enantiomers undergo catalytic alkylation to give constitutional isomers. Since both enantiomers are consumed simultaneously, as the reaction proceeds, the amount of slow enantiomer (relative to the unreacted fast enantiomer) does not increase. Therefore, product ee remains high, even at relatively high conversions.

3
Catalytic RCM and Uncatalyzed Diastereoselective Alkylation of Unsaturated Heterocycles

The availability of non-racemic oxepins through tandem catalytic RCM and Zr-catalyzed kinetic resolution has additional important implications. Optically pure heterocycles that carry a heteroatom within their side chain (cf. (S)-14 in Scheme 3) can be used in stereoselective *uncatalyzed* alkylations. The alcohol, benzyl ether or MEM-ethers derived from (S)-14 readily undergo directed [10] and diastereoselective alkylations when treated with a variety of Grignard reagents [11].

The examples shown in Scheme 5 demonstrate the potential utility of these stereoselective alkylation technologies in synthesis. Thus, preparation of *rac*-14 through catalytic RCM of *rac*-13 (85% yield) and subsequent Zr-catalyzed resolution of the resulting racemic TBS-protected oxepin affords (S)-23 after silyl group deprotection. Diastereoselective alkylation with *n*BuMgBr affords (S)-24

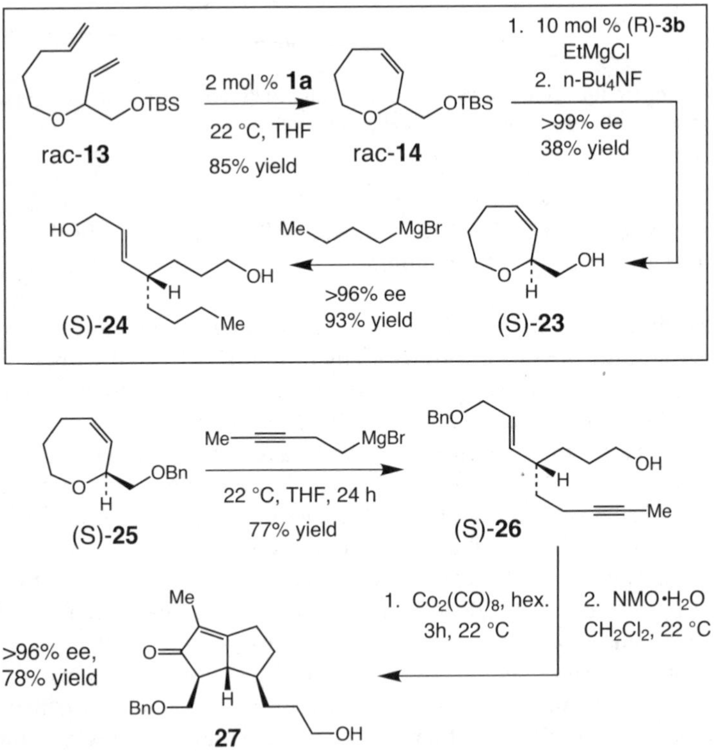

Scheme 5. Chiral medium-ring heterocycles that have been synthesized by catalytic RCM and resolved by the Zr-catalyzed kinetic resolution are subject to diastereoselective alkylations that afford synthetically useful materials in the optically pure form

with >96% ee in 93% yield (>98% trans). As further illustrated in Scheme 5, alkylation of (*S*)-25 with an alkyne-bearing Grignard agent (→(*S*)-26), allows for a subsequent Pauson-Khand cyclization to provide the corresponding bicycle 27 with >98% diastereochemical control and in the optically pure form.

In connection with the facility of these olefin alkylations, it is important to note that the asymmetric Zr-catalyzed alkylations with longer chain Grignard reagents are more sluggish than those involving EtMgCl [4b]. Furthermore, when catalytic alkylation does occur, the corresponding branched products are obtained; that is, with *n*PrMgCl and *n*BuMgCl, *iso*Pr and *sec*Bu addition products are formed, respectively. The uncatalyzed alkylation described here therefore complements the enantioselective Zr-catalyzed protocol.

Catalytic RCM plays two additional roles in enhancing the general utility of the uncatalyzed alkylation represented in Scheme 5:

1. As depicted in Scheme 6, when an alkene-containing Grignard reagent is used, the resulting enantiomerically pure product (e.g., (*S*)-28) can be subjected to 6 mol% 1b to afford the corresponding optically pure carbocycle (*S*)-29 in 65% yield.

2. One of the practical shortcomings of this, and any other, directed reaction [11] is that a Lewis basic site required for reaction may not be desired in the final product. Accordingly, a simple and efficient method for the eventual removal of the directing unit was desired. To address this problem, as shown in Scheme 7, we established that a crotyl unit serves as an effective Lewis basic group to assist stereoselective alkylation (e.g., (*S*)-30→(*S*)-31 in Scheme 7) [12]. Subsequent treatment of (*S*)-31 with 5 mol% Mo catalyst 2, under 1 atm of ethylene, leads to the excision of the directing group and formation of (*S*)-32 in >98% ee and 89% yield.

The catalytic RCM with 31 as substrate (Scheme 7) is significantly more facile when the reaction is carried out under an atmosphere of ethylene, presumably due to the formation of the more active Mo=CH$_2$ system (see below for further details). Thus, after catalytic removal of the directing unit, the chiral unsaturated alcohol (*S*)-32, the formal product of an enantioselective addition of the Grignard reagent to unfunctionalized heterocycle 33, is obtained. An additional in-

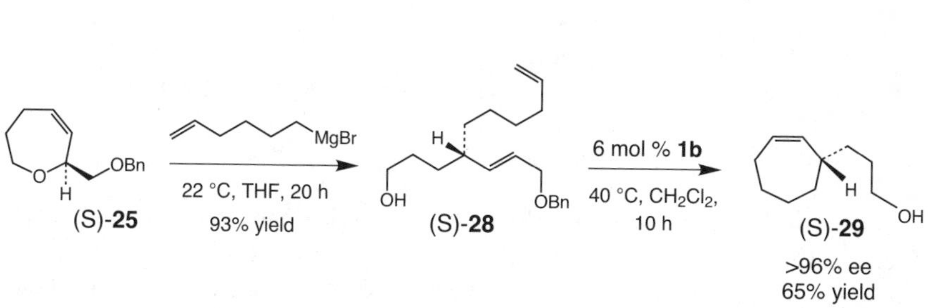

Scheme 6. Chiral medium-ring carbocycles can be prepared by catalytic RCM in conjunction with the directed uncatalyzed alkylation process

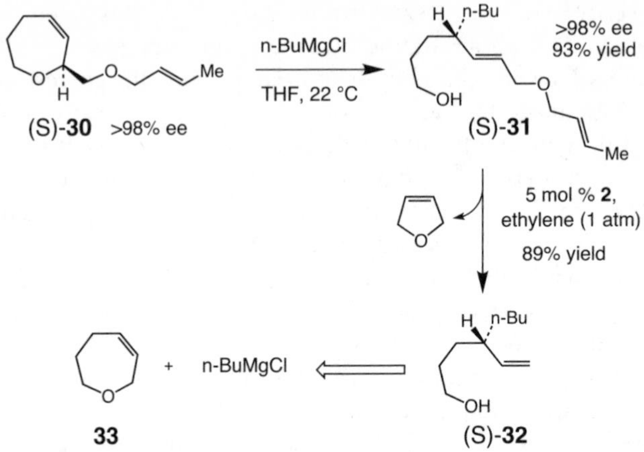

Scheme 7. Uncatalyzed steroselective alkylation proceeds efficiently with a crotyl directing unit, which can be subsequently removed through catalytic RCM

teresting point about this Mo-catalyzed RCM is that – unlike the majority of other examples – it is the released acyclic product, and not the cyclic system (2,5-dihydrofuran in this case), that is the product of interest.

4
Zr-Catalyzed Kinetic Resolution of Allylic Ethers and Ru- and Mo-Catalyzed Synthesis of 2-Substituted Chromenes

Catalytic RCM and another Zr-catalyzed process, the kinetic resolution of cyclic allylic ethers, joined forces in our laboratories in 1995 to constitute a fully-catalytic two-step synthesis of optically pure 2-substituted chromenes. These structural units comprise a critical component of a range of medicinally important agents (see below). Our studies arose from unsuccessful attempts to effect the catalytic kinetic resolution of the corresponding chromenes [13]; a representative example is illustrated in Eq. 3.

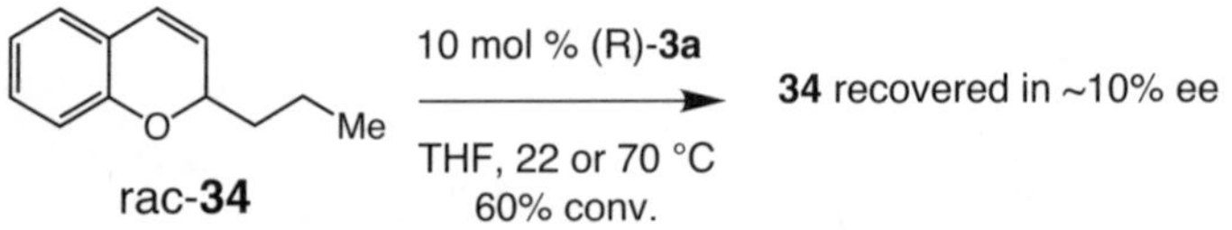

$$(3)$$

Alternatively, as shown in Scheme 8, we envisioned that styrenyl allylic ethers, in the presence of an appropriate catalyst, might undergo a net skeletal rearrangement to yield the desired *isomeric* heterocyclic products [14]. Rearrangement substrates would be synthesized in the non-racemic form by the Zr-catalyzed kinetic resolution [5c].

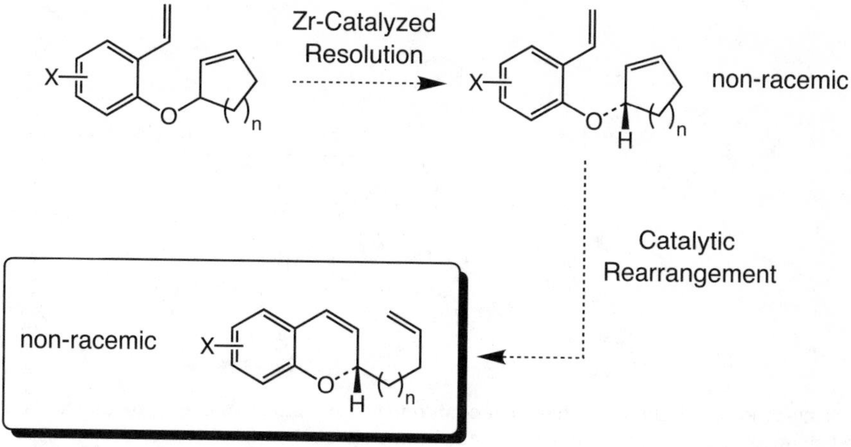

Scheme 8. Zr-catalyzed kinetic resolution of allylic styrenyl ethers may be followed by a Ru- or Mo-catalyzed rearrangement to afford 2-substituted chromenes

4.1
Catalytic Ru-Catalyzed Rearrangements of Terminal Styrenyl Ethers

We conjectured that the proposed catalytic rearrangement would be efficient based on two principles: (1) We were mindful by studies of Crowe that aromatic alkenes and aliphatic olefins are electronically suitable to undergo cross-metathesis [15]. We envisioned that the intramolecular variant should be especially favored. (2) The general reaction appeared energetically favorable: initial semiempirical calculations (PM3; geometry optimization) indicated that the 2-substituted chromenes are appreciably lower in energy than their parent cyclic ethers. This issue is critical in rearrangements, where products can revert back to starting substrates [16]. What added further significance to these considerations was the fact that previous investigations had demonstrated that various metal-catalyzed metathesis reactions [1] may be governed by thermodynamic factors [17].

As the examples in Scheme 9 illustrate, treatment of a styrenyl ether, such as **35**, with 5 mol% Ru catalyst **1a** under an atmosphere of Ar (14 h) leads to the formation of **36** and **37** in 42% and 41% isolated yield, respectively. When the reaction is performed under an atmosphere of ethylene, **36** is obtained in 91% yield. Furthermore, as exemplified by the conversion of **38a** to **39a**, electronic properties of the aromatic moieties exhibit little influence on the facility of the catalytic heterocycle synthesis. Eight-membered rings are appropriate substrates as well (Scheme 9; **38b**→**39b**).

A plausible mechanism for the Ru-catalyzed process is illustrated in Scheme 10. With terminal styrenes such as **40**, reaction begins regioselectively with the formation of metal-carbene **42** (Scheme 10) [18]. Subsequent rear-

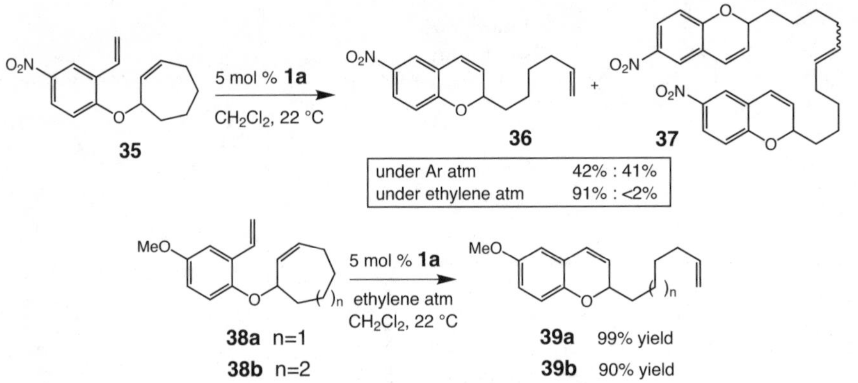

Scheme 9. Ru-catalyzed rearrangement of styrenyl ethers proceeds efficiently under 1 atm ethylene to afford a range of 2-substituted chromenes

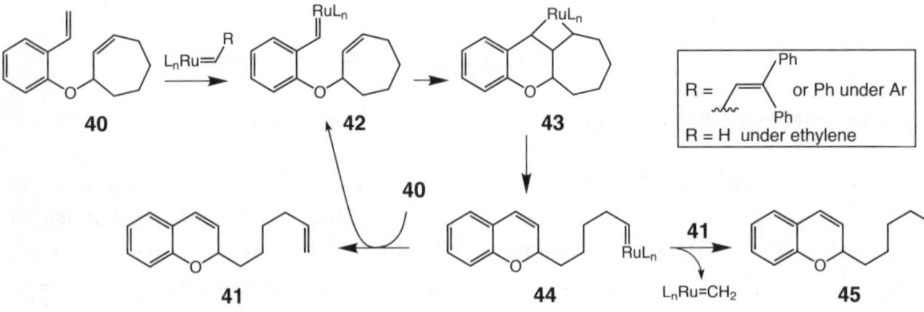

Scheme 10. Mechanism proposed for the Ru-catalyzed rearrangement of terminal styrenyl ethers

rangement, via metallacyclobutane **43**, affords chromene **44**, which can react with a second equivalent of **40** to regenerate **42** and deliver **41**. Late in the process (under Ar atm), as the amount of **41** increases, **44** may react more frequently with the final product monomer (**41**) to afford dimer **45**.

Several factors and observations support the route proposed in Scheme 10: (1) Due to steric factors, the styrenyl alkene is expected to react preferentially (versus the neighboring disubstituted cyclic olefin; see below for further discussion). (2) Involvement of tetracyclic intermediates such as **43** provides a plausible rationale for the reluctance of six-membered ring ethers [**46** in Eq. 4] to participate in the catalytic rearrangement and for the lack of reactivity of cyclopentenyl substrates [**48** in Eq. 5]: because of the attendant angle strain, the generation of the tetracyclic intermediate is not favored. (3) Reactions under ethylene atmosphere inhibit dimer formation, since **44** is intercepted with H_2CCH_2, rather than **41** [19].

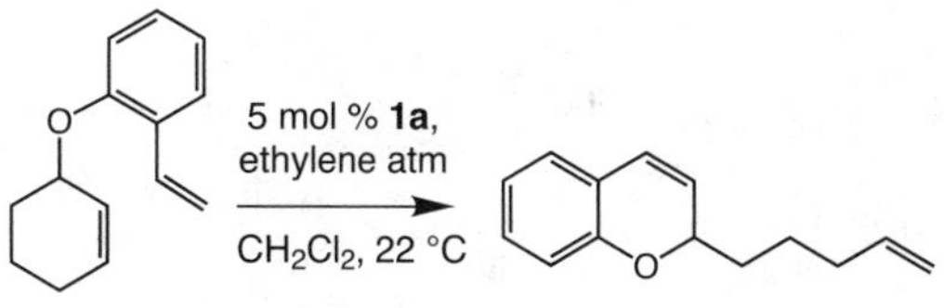

$$(4)$$

46   **47**   35% yield

NO REACTION $$(5)$$

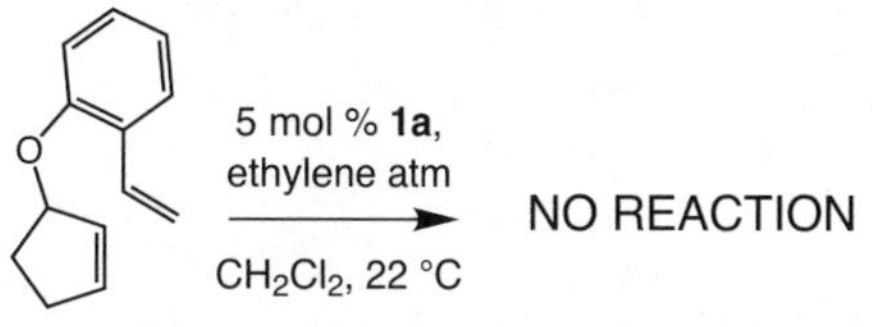

48

4.2
Zr-Catalyzed Resolution and Ru-Catalyzed Reactions of Disubstituted Styrenyl Ethers

As mentioned above, we planned to obtain optically pure styrenyl ethers through Zr-catalyzed kinetic resolution [5]; subsequent metal-catalyzed rearrangement would afford optically pure chromenes. However, as shown in Scheme 11, the recovered starting material (**40**) was obtained with <10% ee (at 60% conversion) upon treatment with 10 mol% (*R*)-(EBTHI)Zr-binol (**3b**) and five equivalents of EtMgCl (70°C, THF). We conjectured that, since the (EBTHI)Zr-catalyzed reaction provides efficient resolution only when asymmetric alkylation occurs at the cyclic alkene site, competitive reaction at the styrenyl terminal olefin renders the resolution process ineffective. Analysis of the ^{1}H NMR spectrum of the unpurified reaction mixture supported this contention. Indeed, as shown in Scheme 11, catalytic resolution of disubstituted styrene **49**

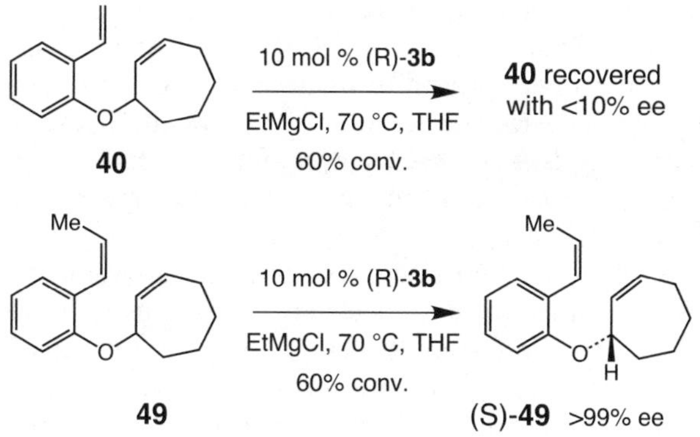

Scheme 11. Zr-catalyzed resolution of disubstituted cycloheptenyl styrenyl ethers, unlike those of terminal styrene derivatives, can be carried out efficiently

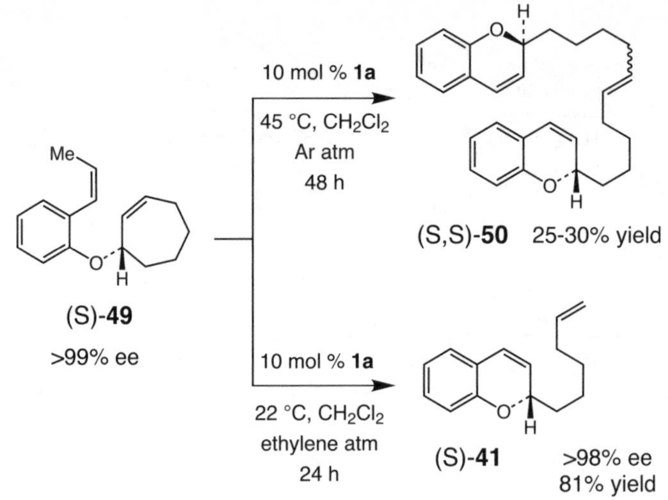

Scheme 12. In the Ru-catalyzed conversion of disubstituted styrenyl ethers to chromenes the presence of ethylene is required for reaction efficiency as well as high yield of monomer formation

proceeded efficiently to afford cycloheptenyl styrene ether (S)-**49** in >99% ee (chiral GLC analysis) and 98% yield (based on percent conversion).

With an effective catalytic resolution of allylic styrene ethers in hand, we focused our attention on the Ru-catalyzed reaction of disubstituted styrenyl ethers (e.g., **49**). When we treated (S)-**49** with 10 mol% **1a** under an atmosphere of Ar (Scheme 12), we found chromene formation to be sluggish: 25–30% of dimer (S,S)-**50** was isolated after 48 h at 45°C, together with substantial amounts of oligomeric materials.

In contrast to the reaction carried under an Ar atm, when (S)-**49** was treated with 10 mol% **1a** under an atmosphere of ethylene (22°C, CH$_2$Cl$_2$, 24 h), (S)-**41** was obtained in 81% isolated yield and >98% ee (Scheme 12). As expected, the use of ethylene atmosphere proved to be necessary for preferential monomer formation (10% of the derived dimer was also generated). These results indicate that the *presence of ethylene is imperative for efficient metal-catalyzed chromene formation as well for processes involving disubstituted styrenyl ethers* (25–30% yield of dimer **50** under argon).

Subsequent mechanistic studies suggested that the abovementioned effect of ethylene on reaction efficiency is connected to a mechanistic divergence that exists for reactions of terminal styrenyl ethers versus those of disubstituted styrene systems [13b]. Whereas with monosubstituted styrenyl substrates the initial site of reaction is the terminal alkene, with disubstituted styrene systems the cyclic π-systems react first. This mechanistic scenario suggests two critical roles for ethylene in the catalytic reactions of disubstituted styrenes:

1. Transformations of the more highly substituted styrenyl ethers are notably more facile under an atmosphere of ethylene due to the presence of the more reactive $L_nRu=CH_2$ (formed by the reaction of **1a** or **1b** with ethylene) [20]. Under an atmosphere of Ar and after the first turnover has transpired, $L_nRu=CHCH_3$ is likely the participating catalyst. When the reaction is performed under ethylene, $LnRu=CHCH_3$ is immediately converted to $LnRu=CH_2$. Reactions of monosubstituted styrenes do not require ethylene to proceed smoothly because, as illustrated in Scheme 10, with this class of starting materials, the more reactive $L_nRu=CH_2$ is formed following the first catalytic cycle.

2. Catalytic reactions of disubstituted styrenyl substrates diminish oligomeric product formation because of the presence of ethylene. That is, if the initial transformation of the Ru-carbene occurs with the "undesired regiochemistry" (e.g., **49**→**52** in contrast to **49**→**51**, Scheme 13), dimerization and oligomerization may predominate, particularly in situations where reclosure of the carbocyclic ring is relatively slow (e.g., cycloheptenyl substrates). In contrast, as illustrated in Scheme 14, in the presence of ethylene atmosphere, the unwanted metal-carbene isomer **52** may rapidly be converted to triene **53**. The resulting triene might then react with $L_nRu=CH_2$ to afford metal-carbene **51** and eventually chromene **41**.

To examine the validity of the above mechanistic paradigm, an authentic sample of triene **53** was prepared and treated with 5 mol% Ru complex **1b** (CH_2Cl_2, 22°C, 24 h). As depicted in Scheme 15, when catalytic RCM is carried out under an atmosphere of Ar, oligomeric products are generated. In contrast, when the reaction is performed in the presence of ethylene, **41** is obtained in 83% isolated yield.

Based on the above principles, cyclopentenyl substrates that bear a disubstituted styrenyl ether should readily afford the desired chromenes by the catalytic

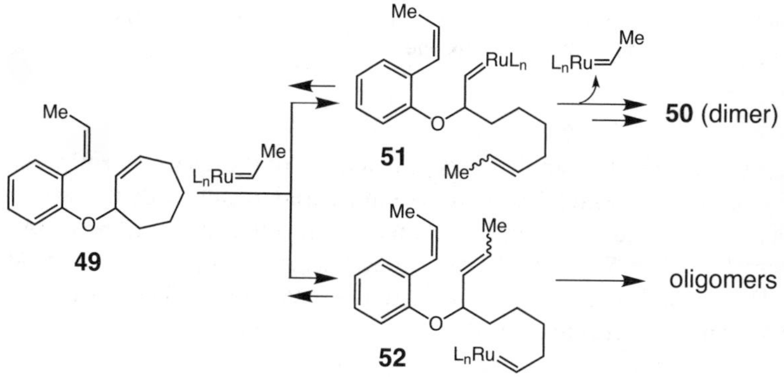

Scheme 13. Reaction pathway for the Ru-catalyzed reactions of disubstituted styrenyl ethers in the absence of ethylene atmosphere

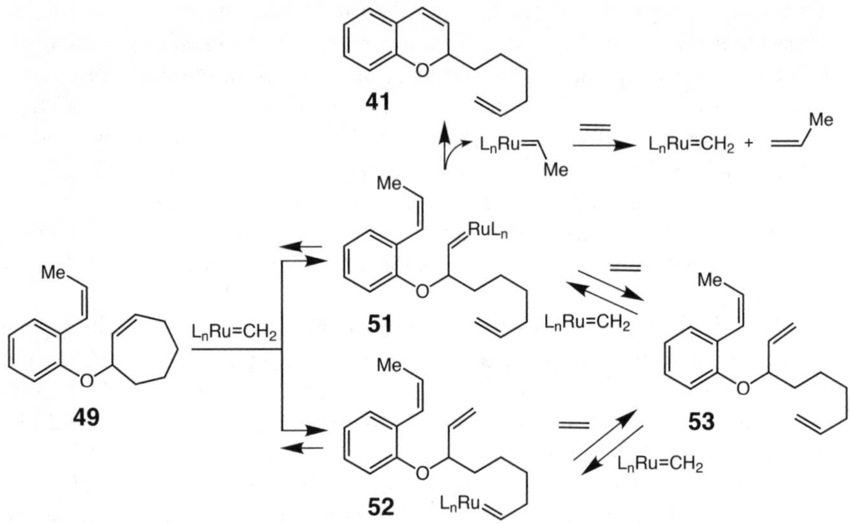

Scheme 14. Reaction pathway for the Ru-catalyzed reactions of disubstituted styrenyl ethers in the presence of ethylene atmosphere

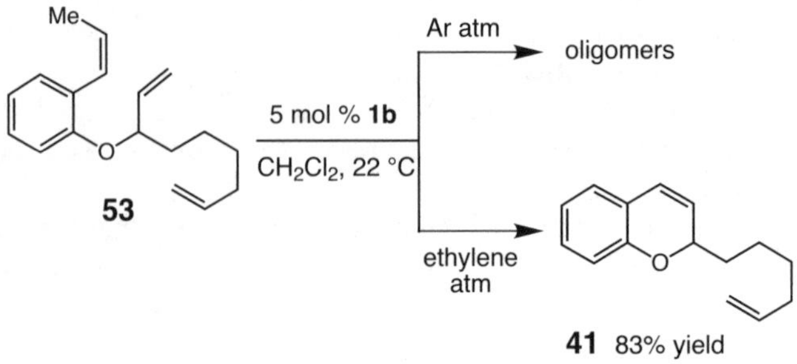

Scheme 15. Reaction pathway for the Ru-catalyzed reactions of disubstituted styrenyl ethers is strongly influenced by ethylene atmosphere

process (in contrast to complete lack of reactivity of the derived terminal styrene **48**, Eq 5). Since the metal-carbene complex first reacts with the carbocyclic alkene, the strained tetracyclic intermediate, formed in reactions of terminal styrenes (cf. **43** in Scheme 10), can be avoided. Indeed, treatment of **54** with 10 mol% **1b** (CH_2Cl_2, 22°C, 50 h) leads to the formation of **55** in 78% yield after silica gel chromatography (Eq. 6).

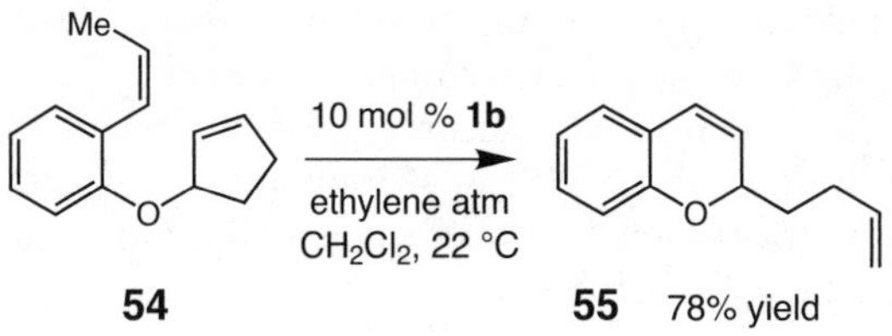

$$(6)$$

54

55 78% yield

[compare to Eq (5)]

56

47 ~15%

56

57

58

59

47 ~15%

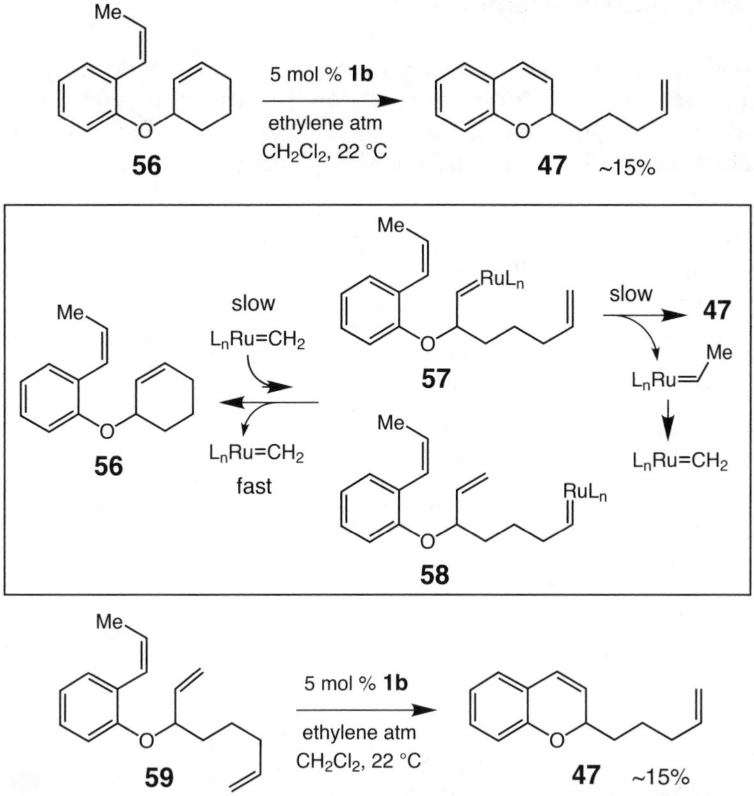

Scheme 16. Reaction pathway for the Ru-catalyzed reactions of cyclohexenyl disubstituted styrenyl ethers is not influenced by the presence of ethylene

The corresponding cyclohexenyl system **56** (Scheme 16) remains relatively unreactive, however, even when the reaction is performed under an ethylene atmosphere: after 24 h (10 mol% **1b**, 1 atm ethylene, CH_2Cl_2), only 10–20% of chromene **47** is obtained. This persistent lack of reactivity is presumably because: (1) the relatively strain-free six-membered ring is less prone (relative to cyclopentenyl and cycloheptenyl structures) to react with $L_nRu=CH_2$ [21], and (2) in case ring rupture does occur with the proper regiocontrol to afford **57** (versus **58**, Scheme 16), reaction with the neighboring terminal alkene (leading back to **56**) should be kinetically more favored (in comparison to reaction with the disubstituted styrene olefin to deliver **47**). Accordingly, when triene **59** is subjected to the reaction conditions, only 15–20% of **47** is obtained (80–85% recovered starting material; 24 h). (See below for an effective method that allows for efficient reactions of cyclohexenyl substrates.)

4.3
Reactions of Functionalized Styrenyl Ethers

An advantage of the metal-catalyzed conversion of styrenyl ethers to chromenes is that control of relative stereochemistry on the carbocyclic substrate before catalytic synthesis of chromenes can lead to the formation of various *functionalized* heterocycles that are diastereomerically pure (Table 1).

Table 1. Synthesis of functionalized chromenes through Ru- and Mo-catalyzed reactions of styrenyl ethers[a]

entry	substrate	product	R		yield(%) Ru cat.[b]	yield(%) Mo cat.[b]
1	**60**	**61**	a	Bn	35	91
			b	TBS	12	93
2	**62**	**63**	a	Bn	36	79
			b	TBS	32	90
3	**64**	**65**	a	H	54	<5
			b	Bn	82	75
			c	TBS	88	97
4	**66**	**67**	a	H	10	<5
			b	Bn	65	98
			c	TBS	55	95

[a] Conditions: 9 mol% **1b**, CH$_2$Cl$_2$, 22°C, ethylene (1 atm), 36 h; 10 mol% **2**, C$_6$H$_6$, 22°C, ethylene (1 atm), 24 h.
[b] Isolated yields after silica gel chromatography.

With the more functionalized substrates, chromene formation typically occurs more efficiently when Mo complex 2 is used instead of Ru-based catalysts 1a or 1b. An impressive example is shown in entry 2 of Table 1; *with 2 as the catalyst, even the relatively unreactive cyclohexenyl substrates such as 62a and 62b are converted to the derived chromenes in excellent yields.* Exceptions to this trend are found in the reactions of 64a and 64b. Particularly striking is the process involving 64a, where the Ru complex 1b affords 65a in 54% yield; in contrast, little or no product is obtained with 2 as the catalyst. The lower activity of 2 is perhaps due to the higher Lewis acidic nature of the Mo center, leading to catalyst inactivation through chelation with the Lewis basic alcohol and the adjacent phenoxy oxygen [22]. It is also possible that the hydroxyl group causes the protonation of Mo-O, Mo=N or Mo=C, leading to partial or complete inactivation of the catalyst. The presence of an unprotected hydroxyl function, such as those shown in entries 3 and 4, appears to cause a notable reduction in reaction efficiency (mass balance remains >95%).

4.4
Zr-Catalyzed Kinetic Resolution of Functionalized Styrenyl Ethers

As was mentioned previously, certain disubstituted styrene ethers can be efficiently resolved through the Zr-catalyzed kinetic resolution. As illustrated in Eq. 7, optically pure cycloheptenyl ether 64c is obtained by the Zr-catalyzed process. The successful catalytic resolution makes the parent alcohol and the derived benzyl ether derivatives 64a and 64b accessible in the optically pure form as well. However, this approach cannot be successfully applied to all the substrates shown in Table 1. For example, under identical conditions, cyclopentenyl susbstrate 60b is recovered in only 52% ee after 60% conversion. Cycloheptenyl substrates shown in entry 4 undergo significant decomposition under the Zr-catalyzed carbomagnesation conditions. These observations indicate that future work should perhaps be directed towards the development of a *chiral* metathesis catalyst that effects the chromene formation and resolves the two styrene ether enantiomers simultaneously.

$$
\text{rac-64c} \xrightarrow[\substack{\text{EtMgCl, 70 °C, THF} \\ \text{65\% conv.}}]{\text{10 mol \% (R)-3b}} \text{(R,R)-64c} \quad >98\% \text{ ee} \tag{7}
$$

4.5
Application to the First Enantioselective Total Synthesis of the Antihypertensive Agent (S,R,R,R)-Nebivolol

To examine and challenge the utility of the two-step catalytic resolution-chromene synthesis process in synthesis [23], we undertook a convergent and enantioselective total synthesis of the potent antihypertensive agent (S,R,R,R)-nebivolol (68) [24]. As illustrated in Scheme 17, the two key fragments (R,S)-71 and (S,S)-73, which were subsequently joined to afford the target molecule, were prepared in the optically pure form by the technology discussed above. Importantly, efficient and selective methods were established for the modification of the chromene alkenyl side chain. These studies therefore have allowed us to enhance the utility of the initial methodological investigations: they demonstrate that, although the carbocyclic system may be used as the framework for the Zr- and the Mo-catalyzed reactions, the resulting 2-substituted chromene can be functionalized in a variety of manners to afford a multitude of chiral non-racemic heterocycles [25].

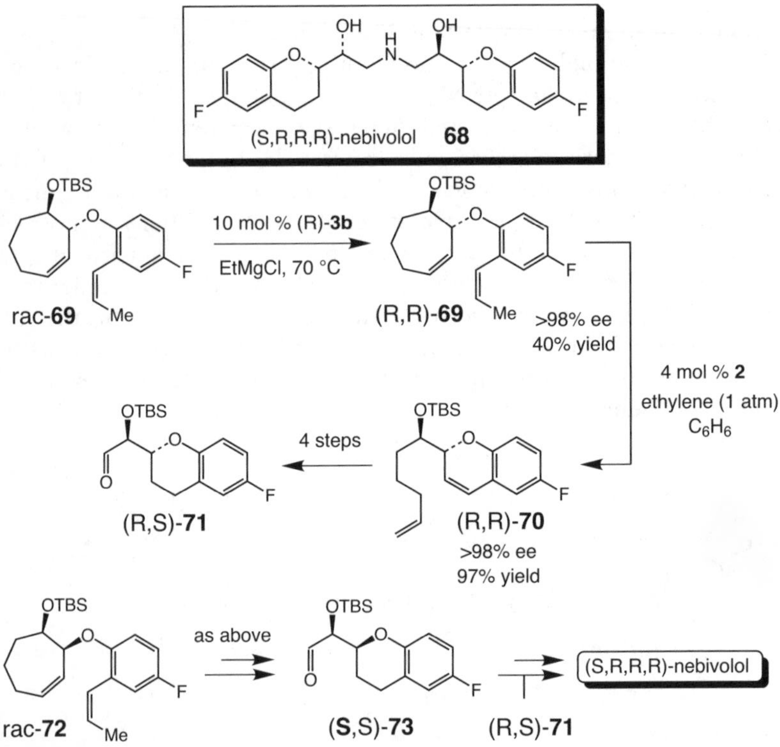

Scheme 17. The tandem Zr-catalyzed kinetic resolution and Mo-catalyzed conversion of styrenyl ethers to chromenes is used in the first convergent and enantioselective total synthesis of the antihypertensive agent (S,R,R,R)-nebivolol

5
Mo-Catalyzed Stereoselective Synthesis of Macrocyclic Trisubstituted Alkenes

5.1
Studies on Sch 38516 Aglycon

One of our earliest forays into the use of catalytic RCM was concieved in 1994, during the revision of our plans towards the total synthesis of the antifungal agent Sch 38516 (**74**; also known as fluvirucin B$_1$) [26]. We were required to re-evaluate our overall synthesis strategy, since our initial approach – one that was along the more established lines (Wittig olefination) – in constructing the 14-membered ring macrocycle had failed. We therefore conjectured that stereoselective olefin synthesis through macrocyclization by catalytic RCM could be followed by Pd-catalyzed hydrogenation of the olefinic site. It was our hope that this two-step and fully-catalytic process would provide an effective solution to one of the more challenging problems in this total synthesis: control of the remote C6 stereogenic center. Our planning was also founded on initial molecular modeling (on the TBS-protected ring structure), indicating that catalytic hydrogenation of the resident olefin would benefit from selective peripheral addition [27] to deliver the desired stereoisomer.

Sch 38516 **74**
(fluvirucin B$_1$)

Thus, in seeking an alternative and more efficient scenario to the formation of macrolactam **75** (Scheme 18), we considered the possibility of using catalytic RCM to reach the desired trisubstituted macrocyclic alkene. At the time, synthesis of an 8- and a 13-membered heterocycle through the use of such catalytic procedures had been disclosed by Martin [28] and Pandit [29]. In the studies reported by Martin, a disubstituted alkene was efficiently prepared with Schrock's Mo catalyst (**2**), whereas in the case of Pandit, Grubbs's complex **1a** was used. In both instances, however, the starting diene benefited from a rigid cyclic framework. Another report by Martin [30] illustrated that efficient synthesis of 12-membered rings from conformationally mobile aliphatic starting materials could be difficult under standard conditions. Our consternation about the importance of the conformational rigidity of the precursor diene on the facility of catalytic RCM of medium and large rings was exacerbated by Grubbs's report on the influence of structural scaffolds on the facility of eight-membered ring construction promoted by **1a** [6a].

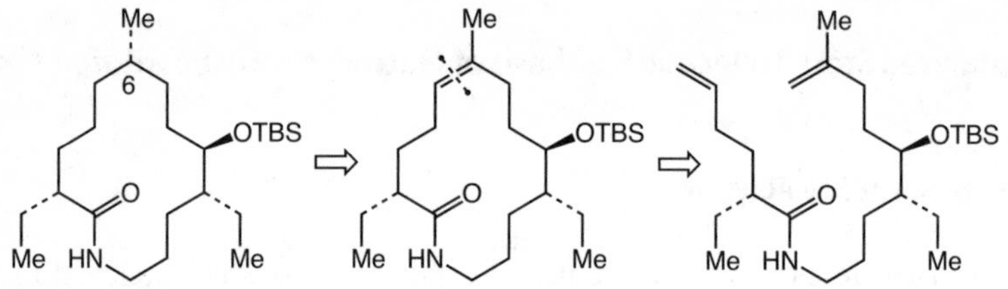

Scheme 18. Retrosynthesis for the construction of the 14-membered macrolactam ring of Sch 38516 (fluvirucin B$_1$)

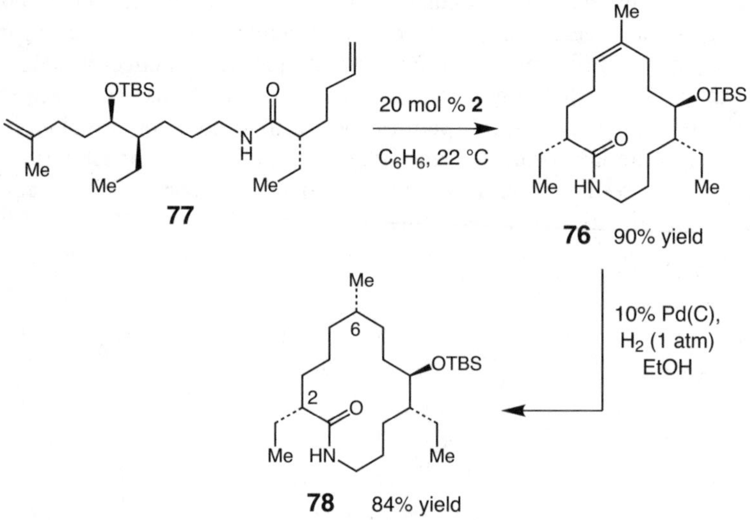

Scheme 19. Mo-catalyzed RCM in tandem with catalytic hydrogenation allows the control of C2–C6 relative stereochemistry in the macrolactam region of Sch 38516

Because of the above concerns, the prospects for an efficient and reliable catalytic ring closure of **77** to afford **76** did not appear certain (Scheme 18). We were aware that the presence of stereogenic centers have been reported to infer appreciable degrees of structural rigidity on a number of occasions [31, 32]. But we also kept in mind that our plan required facile formation of a *tri*substituted macrocyclic olefin: substrate oligomerization, catalyst decomposition and conformational freedom would not be as readily overcome when the desired product is a slower-forming and highly substituted alkene [33].

To explore the validity of the ring formation strategy, we drew up plans and carried out an enantioselective synthesis of diene **77** (Scheme 19). Initially, based on previous studies on similar medium- and large-ring syntheses, we surmised that elevated temperatures would be necessary for effective macrocyclization. We therefore excluded ambient temperature conditions from our preliminary investigations. Accordingly, we established that, when 77 is treated with 20 mol% Mo catalyst 2 in benzene (0.01 M) and the mixture is heated to 50°C (12 h), macrolactam 76 is obtained in 60–65% isolated yield as a single olefin stereoisomer (Scheme 19).

We later determined that with lower catalyst loadings (15 mol% 2), reaction efficiency suffers (<50% conversion). When Ru complex 1b was used (20 mol%, benzene, 80°C), 5–10% dimer derived from reaction of the terminal olefins was formed as the only product. Continued investigation of the catalytic macrocyclization indicated that, *with freshly prepared and recrystallized Mo catalyst, the metathesis process occurs smoothly at 22°C to afford 2 in 90% isolated yield after only 4 h.* With 40 mol% 2, the yield improved to 97%, and less than 20 mol% catalyst gave notably lower conversions and yields.

With the Mo-catalyzed macrolactamization secured, we turned our attention to the problem of C2-C6 remote stereochemical control. Catalytic hydrogenation of unsaturated cyclic amide 76, in the presence of 10% Pd(C), resulted in the formation of 78 in 84% yield and with >98% diastereoselectivity.

5.2
Origin of the Facile Mo-Catalyzed Macrocyclization

To gain insight into this remarkably efficient cyclization reaction (77→76), and to determine the extent to which existing stereogenic centers preorganize diene precursor 77, we examined the catalytic RCM of 79 and 82. When 79 was subjected to 25 mol% 2 (50°C, 18 h), <2% 80 was formed (Scheme 20). Instead, dimer 81 was obtained in 52% yield (3:1 mixture of olefin isomers; identity of major product not determined). When diene 82 was treated with identical conditions, macrolactam 83 was obtained in 41% isolated yield [33] along with 20% of the

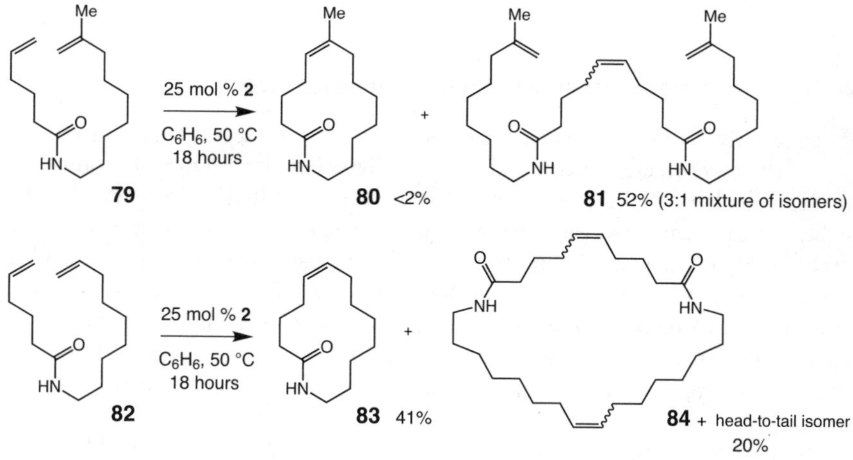

Scheme 20. In the absence of stereogenic centers, catalytic RCM may suffer significantly

derived 28-membered ring cyclic dimer **84** (presumably as a mixture of head-to-head and head-to-tail isomers); <10% of the acyclic dimer was detected. *These data clearly illustrate that with the slower forming trisubstituted olefin, the presence of stereogenic centers is required for facile ring closure* (cf. **77→76**, Scheme 19), whereas when reacting alkenes are terminal olefins, formation of the 14-membered ring is more favored (41% yield of **83**). Our results lend further credence to Grubbs's suggestion [6] that, with conformationally mobile diene structures, some form of structural restraint is needed for facile ring-closing metathesis.

Considering the facility with which dimerization products **81** and **84** are obtained, we reasoned that, in catalytic ring closure of **77**, the derived dimer is perhaps initially formed as well. If the metathesis process is reversible [17b], such adducts may subsequently be converted to the desired macrocycle **76**. To examine the validity of this paradigm, diene **77** was dimerized (→**85**) by treatment with Ru catalyst **1b**. When **85** was treated with 22 mol% **2** (after pretreatment with ethylene to ensure formation of the active complex), 50–55% conversion to macrolactam **76** was detected within 7 h by 400 MHz ^{1}H NMR analysis (Eq. 8). When **76** was subjected to the same reaction conditions, <2% of any of the acyclic products was detected. Although we do not as yet have a positive proof that **85** is formed in cyclization of **77**, this observation suggests that if dimerization were to occur, the material can be readily converted to the desired macrolactam, which is kinetically immune to cleavage.

$$(8)$$

5.3
Mo-Catalyzed Macrocyclization of Carbohydrate-Bearing Diene

Subsequently, in the course of the total synthesis, we established that glycosylation of the macrolactam alcohol or various silyl ethers by a range of protocols is not feasible. Repeated initiatives along these lines resulted in complete decomposition of the carbohydrate systems and low recovery of the macrocyclic substrate (<15%). The principal reason for such unsatisfactory outcomes proved to be the notorious lack of solubility of the parent alcohol of **76** in a wide variety of solvents; the 14-membered alcohol is sparingly soluble in methanol at ambient temperatures. Clearly, our initial efforts towards the use of catalytic RCM in constructing the macrolactam segment of Sch 38516 required additional investigation.

The above observations implied that a more viable strategy would be to achieve stereoselective glycosylation of the more readily soluble acyclic diene to

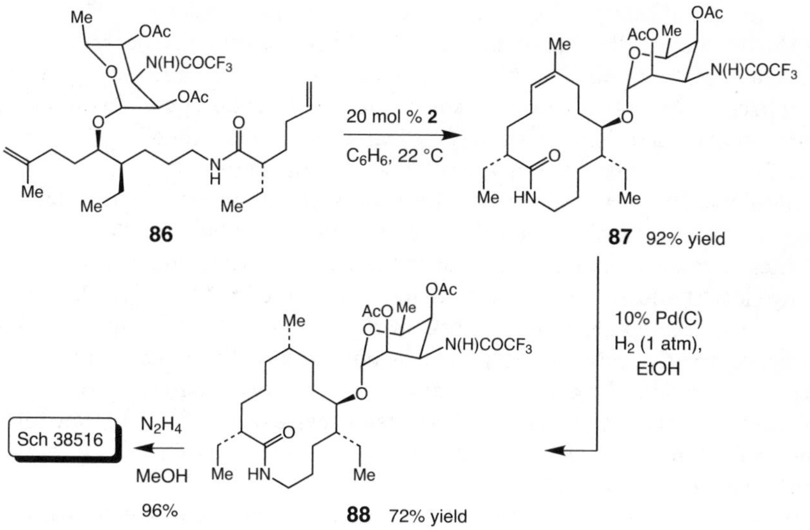

Scheme 21. Mo-catalyzed RCM is carried out on the fully functionalized diene; subsequent hydrogenation delivers the protected form of Sch 38516

obtain **86** (Scheme 21), which may then be induced to undergo catalytic RCM. Importantly, the success of the proposed route depended on whether the Lewis acidic metathesis catalyst (**2**) would remain operative in the presence of additional Lewis basic heteroatoms carried by the pendant carbohydrate moiety.

As depicted in Scheme 21, subjection of **86** with 20 mol% of freshly prepared **2** after 4 h at 22°C indeed afforded **87** in 92% yield after silica gel chromatography (>98% Z). Stereocontrolled hydrogenation of the trisubstituted olefin (72% yield) and removal of the acetate and trifluroacetate groups, effected by subjection of the hydrogenated adduct with hydrazine in MeOH, delivered Sch 38516 (**1**) in 96% yield to complete the total synthesis.

6
Conclusions and Outlook

The chemistry described in this review article demonstrates the impressive positive influence that catalytic RCM has had on our research in connection to the development of other catalytic and enantioselective C-C bond forming reactions. There is no doubt that in the absence of pioneering work by Schrock and Grubbs, the Zr-catalyzed alkylation and kinetic resolution would be of less utility in synthesis. The number of unsaturated heterocyclic and carbocyclic substrates available for Zr-catalyzed asymmetric carbomagnesation would be far more limited without catalytic RCM.

The enantioselective total syntheses of the antihypertensive agent (S,R,R,R)-nebivolol and the antigungal agent Sch 38516 (fluvirucin B$_1$) illustrate the positive impact that catalytic RCM may have on a total synthesis effort. The efficient and facile catalytic RCM of dienes **77** and **86** illustrate that this macrocyclization procedure offers a mild and stereoselective route to the synthesis of highly functionalized macrocycles bearing trisubstituted olefins. Subsequent to our preliminary reports [26], elegant studies by Danishefsky [34], Nicolaou [35], Fuchs [36], Fürstner [37], and Schinzer [38] have provided additional and impressive testimony in support of the utility of this protocol in the preparation of macrocyclic disubstituted alkenes (with **1a** or **1b** as catalyst) [39].

There is much more left to be done in the area of catalytic RCM. It is likely that many elegant and creative uses of catalytic RCM are in the making. Judging by the related developments in recent years, it is also likely that catalytic RCM will influence positively the development of numerous other ongoing metal-catalyzed or uncatalyzed reactions. The advent and utility of complexes **1a**, **1b** and **2** will undoubtedly inspire organic chemists to devise new and useful transformations, where these transition metal systems are effectively utilized (e.g., styrenyl ether rearrangements) [40].

Perhaps the most compelling research objective in this area will involve the development of a chiral metathesis catalyst that effects C-C bond formation efficiently and with excellent levels of enantioselectivity [41]. In such a case, all the reactions discussed herein, in addition to those expertly developed in other laboratories [40], will become subject to asymmetric catalysis. Such a development should prove to have an enormous impact on the field of inorganic, organometallic and synthetic organic chemistry.

Acknowledgements
National Institutes of Health (GM-47480) and the National Science Foundation (CHE-9257580 and CHE-9632278) generously supported the research described in this article. Additional assistance from Johnson and Johnson, Pfizer, Eli Lilly, Zeneca, Glaxo and Monsanto is gratefully acknowledged as well. I am forever grateful to my talented coworkers and colleagues Zhongmin Xu, James P. Morken, Ahmad F. Houri, Mary T. Didiuk, Michael S. Visser, Joseph P. A. Harrity, Charles W. Johannes, Nicola M. Heron, Jeffrey A. Adams, Daniel S. La, Dustin R. Cefalo, Gabriel A. Weatherhead and John D. Gleason; the research detailed above is nothing but the result of their dedication, perseverance and ingenuity. This article is dedicated to my mentor Professor Thomas J. Katz (Columbia University), in celebration and recognition of his ground-breaking early contributions to the field of olefin metathesis.

7
References and Footnotes

1. For recent reviews on olefin metathesis in organic synthesis, see: (a) Grubbs RH, Miller SJ, Fu GC (1995) Acc Chem Res 28:446 and references cited therein. (b) Schmalz H-G (1995) Angew Chem, Int Ed Engl 107:1833 and references cited therein. (c) Schuster M, Blechert S (1997) Angew Chem, Int Ed Engl 36:2036. For initial pioneering studies on olefin metathesis, see: (d) Katz TJ, Lee SJ, Acton N (1976) Tetrahedron Lett 4247. (e) Katz TJ, Acton N (1976) Tetrahedron Lett 4241. (f) Katz TJ, McGinnis J, Altus C (1976)

J Am Chem Soc 98:606. (g) Katz TJ (1977) Adv Organomet Chem 16:283. (h) Tsuji J, Hashiguchi S (1980) Tetrahedron Lett 21: 2955

2. (a) Fu GC, Grubbs RH (1992) J Am Chem Soc 114:7324. (b) Fu GC, Grubbs RH (1993) J Am Chem Soc 115:3800. (c) Wu Z, Nguyen ST, Grubbs RH, Ziller JW (1995) J Am Chem Soc 117:5503

3. (a) Schrock RR, Murdzek JS, Bazan GC, Robbins J, DiMare M, O'Regan M (1990) J Am Chem Soc 112:3875. (b) Bazan GC, Schrock RR, Cho H-N, Gibson VC (1991) Macromolecules 24:4495

4. (a) Morken JP, Didiuk MT, Hoveyda AH (1993) J Am Chem Soc 115:6697. (b) Didiuk MT, Johannes CW, Morken JP (1995) J Am Chem Soc 117:7097. For a recent review, see: Hoveyda AH, Morken JP (1996) Angew Chem, Int Ed Engl 35:1262

5. (a) Morken JP, Didiuk MT, Visser MS, Hoveyda AH (1994) J Am Chem Soc 116:3123. (b) Visser MS, Hoveyda AH (1995) Tetrahedron 51:4383. (c) Visser MS, Harrity JPA, Hoveyda AH (1996) J Am Chem Soc 118:3779. (d) Visser MS, Heron NM, Didiuk MT, Sagal JF, Hoveyda AH (1996) J Am Chem Soc 118:4291

6. (a) Miller SJ, Kim S, Chen Z, Grubbs RH (1995) J Am Chem Soc 117:2108. (b) Miller SJ, Grubbs RH (1995) J Am Chem Soc 117:5855. (c) Houri AF, Xu Z, Cogan DA, Hoveyda AH (1995) J Am Chem Soc 117:2943

7. For a discussion of the Thorpe-Ingold effect, see: Eliel E, Wilen SH, (1994) Stereochemistry of organic compounds, 1st edn. Wiley Interscience, New York, p 682

8. For a related recent report, see: (a) Linderman RJ, Siedlecki J, O'Neill SA, Sun H (1997) J Am Chem Soc 119:6919. (b) Furstner A, Muller T (1997) Synlett 1010

9. For two examples of this type of resolution (the first two involve transition metal catalysis), see: (a) Hayashi T, Yamamoto M (1987) Chem Lett 177. (b) Martin SF, Spaller MR, Liras S, Hartmann B (1994) J Am Chem Soc 116:4493. (c) Vedejs E, Chen X (1997) J Am Chem Soc 119:2584

10. For a review of directed reactions, see: Hoveyda AH, Evans DA, Fu GC (1993) Chem Rev 93:1307

11. Heron NM, Adams JA, Hoveyda AH (1997) J Am Chem Soc 119:6205

12. Heron NM, Adams JA, Hoveyda AH (1998) manuscript in preparation

13. (a) Harrity JPA, Visser MS, Gleason JD, Hoveyda AH (1997) J Am Chem Soc 119:1488. (b) Harrity JPA, La DS, Cefalo DR, Visser MS, Hoveyda AH (1998) J Am Chem Soc 120:2343

14. Unlike what is typically observed with Ru-catalyzed ring-closing metatheses (ref. 5a,b), the products of reactions reported herein are isomers of the starting materials; the Ru-catalyzed reactions thus constitute a rearrangement.

15. (a) Crowe WE, Zhang ZJ (1993) J Am Chem Soc 115:10998. (b) Crowe WE, Goldberg DR (1995) J Am Chem Soc 117:5162. (c) Crowe WE, Goldberg DR, Zhang ZJ (1996) Tetrahedron Lett 37:2117

16. Catalytic transformations of both terminal and disubstituted styrenyl ethers will be discussed in this article. In the former case, since the starting material and the product are isomeric, the Ru-catalyzed process constitutes a catalytic rearrangement.

17. (a) Ref. 6a. (b) Marsella MJ, Maynard HD, Grubbs RH (1997) Angew Chem, Int Ed Engl 36:1101 and references cited therein

18. For a recent report on the mechanism of the Ru-catalyzed ring-closing metathesis, see: Dias EL, Nguyen ST, Grubbs RH (1997) J Am Chem Soc. 119:3887

19. Treatment of **45** with 5 mol% **1a** under an atmosphere of ethylene leads to 50% conversion to **41** (12 h; 400 MHz ^{1}H NMR analysis). Therefore, in the presence of ethylene, even if dimer is formed, it can be readily converted to the monomeric form with reasonable efficiency.

20. Schwab P, Grubbs RH, Ziller JW (1996) J Am Chem Soc 118:100

21. For a study on the ring-opening metathesis of cyclohexene, see: Patton PA, Lillya CP, McCarthy TJ (1986) Macromolecules 19:1266

22. For a hydroxyl-directed olefination of ketones with **2** as catalyst, see: Fujimura O, Fu GC, Rothemund PWK, Grubbs RH (1995) J Am Chem Soc 117:2355

23. For the Mn-catalyzed kinetic resolution of 2,2-disubstituted chromenes, see: Vander Velde SL, Jacobsen EN (1995) J Org Chem 60:5380

24. Van Lommen G, De Bruyn M, Schroven M (1990) J Pharm Belg 45:355 and references cited therein

25. Johannes CW, Visser MS, Weatherhead GA, Hoveyda AH (1998) J Am Chem Soc in press

26. Xu Z, Johannes CW, Houri AF, La DS, Cogan DA, Hofilena GE, Hoveyda AH (1997) J Am Chem Soc 119:10302 and references cited therein

27. (a) Corey EJ, Hopkins PB, Kim S, Yoo S-E, Nambia KP, Falck JR (1979) J Am Chem Soc 101:7131. (b) Still WC, Galynker I (1981) Tetrahedron 37:3981. (c) Schreiber SL, Santini C (1984) J Am Chem Soc 106:4038. (e) Neeland EG, Ounsworth JP, Sims RJ, Weiler L (1994) J Org Chem 58:7383. (f) Evans DA, Ratz AM, Huff BE, Sheppard GS (1995) J Am Chem Soc 117:3448

28. (a) Martin SF, Liao Y, Wong Y, Rein T (1994) Tetrahedron Lett 35:691

29. Borer BC, Deerenberg S, Bieraugel H, Pandit UK (1994) Tetrahedron Lett 35:3191

30. Martin SF, Liao Y, Chen H-J, Patzel M, Ramser MN (1994) Tetrahedron Lett 35:6005

31. For a classic example, see: Woodward RB, et. al. (1981) J Am Chem Soc 103:3213

32. Recent studies by Grubbs on the effect of stereochemistry on synthesis of macrocyclic peptides (disubstituted alkenes) by ring closing metathesis were not disclosed at the time of our planning. See: (a) Ref. 6a. (b) Miller SJ, Blackwell H, Grubbs RH (1996) J Am Chem Soc 118:9606

33. As recently demonstrated by Furstner and Langemann, higher yields of the disubstituted olefin **83** can be obtained under high dilution conditions using **1a** as the catalyst. Our experiments clearly illustrate that synthesis of trisubstituted macrocyclic alkenes is more complicated than that of their disubstituted analogues. See: Furstner A, Langemann K (1996) J Org Chem 61:3942

34. (a) Bertinato P, Sorensen EJ, Meng D, Danishefsky SJ (1996) J Org Chem 61:8000. (b) Meng D, Su D-S, Balog A, Bertinato P, Sorensen EJ, Danishefsky SJ, Zheng Z-H, Chou T-C, He L, Horwitz SB (1997) J Am Chem Soc 119:2733. (c) Meng D, Bertinato P, Balog A, Su D-S, Kamenecka T, Sorensen EJ, Danishefsky SJ (1997) J Am Chem Soc 119:10073

35. (a) Nicolaou KC, He Y, Vourloumis D, Vallberg H, Yang Z (1996) Angew Chem, Int Ed Engl 35:2399. (b) Yang Z, He Y, Vourloumis D, Vallberg H, Nicolaou KC (1997) Angew Chem, Int Ed Engl 36:166. (c) Nicolaou KC, Sarabia F, Ninkovic S, Yang Z (1997) Angew Chem, Int Ed Engl 36:525. (d) Nicolaou KC, Winssinger N, Pastor J, Ninkovic S, Sarabia F, He Y, Vourloumis D, Yang Z, Li T, Giannakakou P, Hamel E (1997) Nature 387:268

36. Kim SH, Figueroa I, Fuchs PL (1997) Tetrahedron Lett 38:2601

37. (a) Furstner A, Kindler K (1996) Tetrahedron Lett 37:7005 (b) Furstner A, Langemann K (1997) J Am Chem Soc 119:9130. (c) Furstner A, Langemann K (1997) Synthesis 792

38. Schinzer D, Limberg A, Bauer A, Bohm OM, Cordes M (1997) Angew Chem, Int Ed Engl 36:523

39. More recent accounts from the laboratories of Grubbs and Sauvage indicate that the metathesis technology may be readily applied to the synthesis of macrocyclic crown ethers and [2]catenanes. See: (a) Ref. 17b. (b) Mohr B, Weck M, Sauvage J-P, Grubbs RH (1997) Angew Chem, Int Ed Engl 36:1308

40. For recent examples, where **1a** has been used in the development of new reaction methods, see: (a) Clark TD, Ghadiri MR (1995) J Am Chem Soc 117:12364. (b) Tallarico JA, Randall ML, Snapper ML (1996) J Am Chem Soc 38:9196. (c) Kim S-H, Zuercher WJ, Bowden NB, Grubbs RH (1996) J Am Chem Soc 118:1073. (d) Snapper ML, Tallarico JA, Randall ML (1997) J Am Chem Soc 119:1478. (e) Piscopio AD, Miller JF, Koch K (1997) Tetrahedron Lett 38:7143. (f) Kinoshita A, Sakakibara N, Mori M (1997) J Am Chem Soc 119:12388. (g) Crimmins MT, Choy AL (1997) J Org Chem 62:7548

41. For a recent report on Mo-catalyzed enantioselective RCM, see: Alexander JB, La DS, Cefalo DR, Hoveyda AH, Schrock RR (1998) J Am Chem Soc 120:4041 and references cited therein

Enyne Metathesis

Miwako Mori

Enyne metathesis is unique and interesting in synthetic organic chemistry. Since it is difficult to control intermolecular enyne metathesis, this reaction is used as intramolecular enyne metathesis. There are two types of enyne metathesis: one is caused by [2+2] cycloaddition of a multiple bond and transition metal carbene complex, and the other is an oxidative cyclization reaction caused by low-valent transition metals. In these cases, the alkylidene part migrates from alkene to alkyne carbon. Thus, this reaction is called an alkylidene migration reaction or a skeletal reorganization reaction. Many cyclized products having a diene moiety were obtained using intramolecular enyne metathesis. Very recently, intermolecular enyne metathesis has been developed between alkyne and ethylene as novel diene synthesis.

Keywords: Enyne metathesis, Enyne, Metathesis, Carbene complex, [2+2] Cycloaddition

1
Introduction

The metathesis reaction is a powerful strategy in synthetic organic chemistry [1], and it is generally accepted that this reaction is catalyzed by highly efficient transition metal alkylidenes [2, 3]. Intermolecular diene metathesis produces

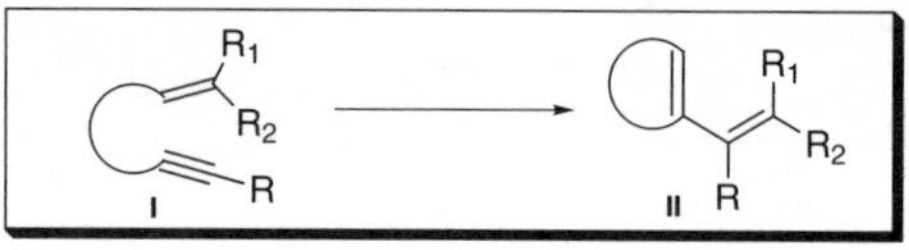

Scheme 1. Enyne metathesis

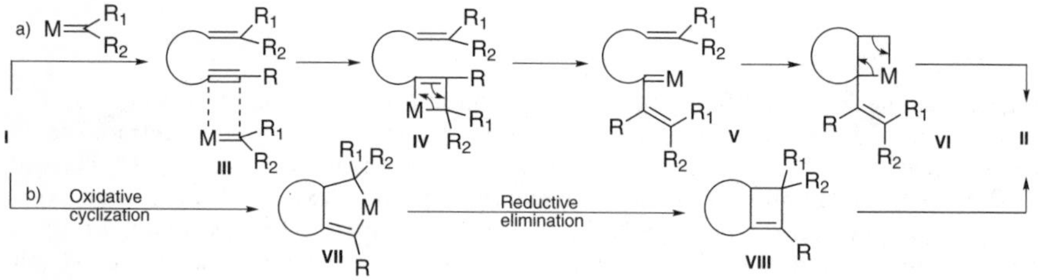

Scheme 2. Reaction proceeding by a low-valent transition metal complex

many olefins [4], and it has usually been used as intramolecular diene metathesis [5]. Enyne metathesis is also quite interesting, and this reaction occurs between alkene and alkyne. Since it is difficult to control this reaction, we also used it as intramolecular enyne metathesis. This reaction is designated as an alkylidene migration reaction from the alkene part to the alkyne carbon (Scheme 1). Thus, enyne metathesis is also called a skeletal reorganization or alkylidene migration.

Enyne metathesis is caused by transition metals. There are two types of enyne metathesis: one is caused by a carbene complex, as is olefin metathesis, via [2+2] cocyclization; and the other type is a reaction that proceeds via oxidative cyclization by a low-valent transition metal complex (Scheme 2).

Enyne metathesis was discovered by Katz [6] in 1985. Katz demonstrated this reaction as a methylene migration reaction using Fischer tungsten carbene complex. The same types of reactions were subsequently reported using Fischer molybdenum and chromium carbene complexes. Recently, very reactive carbene complexes for olefin metathesis were discovered by Grubbs and Schrock. Grubbs synthesized many cyclic compounds using his ruthenium carbene complex. This methodology is now a useful synthetic method for cycloalkenes in the field of natural product synthesis, especially macrocyclic compounds. It has become apparent that the ruthenium carbene complex acts as a catalyst for dienyne- and enyne metathesis. On the other hand, Trost discovered a very interesting palladium-catalyzed enyne metathesis during the course of his study on enyne cyclization. This reaction proceeds through oxidative cyclization by a low-valent metal complex. It is now known that many metals are effective for enyne metathesis.

2
Intramolecular Enyne Metathesis

2.1
Enyne Metathesis Using Fischer Carbene Complexes (W, Mo, Cr)

Katz [6] reported a very interesting methylene migration reaction using a Fischer tungsten carbene complex. The reaction of enyne 1 and 1 mol% of tungsten carbene complex 2a gave methylene migration product 3 in 31% yield (Eq. 1). However, when an equimolar amount of complex 2b was used for this reaction, a different metathesized product 5 was obtained in 50% yield (Eq. 2). This is the first example of intramolecular enyne metathesis. In the former reaction, the real catalyst is methylene tungsten complex 4 (R=R'=H). In this reaction, the alkyne part of enyne 1 reacts with methylene tungsten complex 4 generated from 1 and 2a via [2+2]cycloaddition to form metalacyclobutene 6 (Scheme 3). Retrocycloaddition occurs to give vinyl carbene complex 7, which reacts with the alkene intramolecularly to produce metalacyclobutane 8. Retrocycloaddition occurs to give 3 and methylene tungsten complex 4.

$$(1)$$

$$(2)$$

Following the report by Katz, Hoye [7] reported that enyne 9 having a dimethyl group on the alkene gave metathesis products 10 and 11 using a stoichiometric amount of a Fischer carbene complex (Eq. 3, Table 1).

$$(3)$$

The substituent effects on the alkene were investigated in the reaction of enyne 12 and chromium carbene complex 2c [8]. In the reaction of enyne E-12a having a phenyl group on the alkene with Fischer chromium carbene complex 2c, metathesis product 13a was obtained as a main product along with cyclopropane 14 and cyclobutanone 15 (Eq. 4). The reaction of Z-12a with 2c gave only

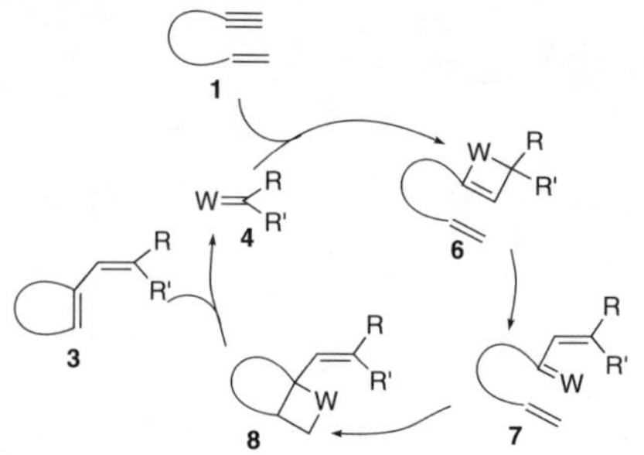

Scheme 3. Reaction mechanism

Table 1. Reaction of enyne **9** with **2a**

M	Yield (%) of **10**	Yield (%) of **11**
Cr	30	–
Mo	28	50
W	6	34

metathesis products **13a** and **16** (Eq. 5). Investigation of the substituent effects on the alkene showed that an electron-withdrawing group accelerates the formation of cyclopropane **14**, while the enyne having an electron-donating group gave the metathesis product as a main product (Table 2).

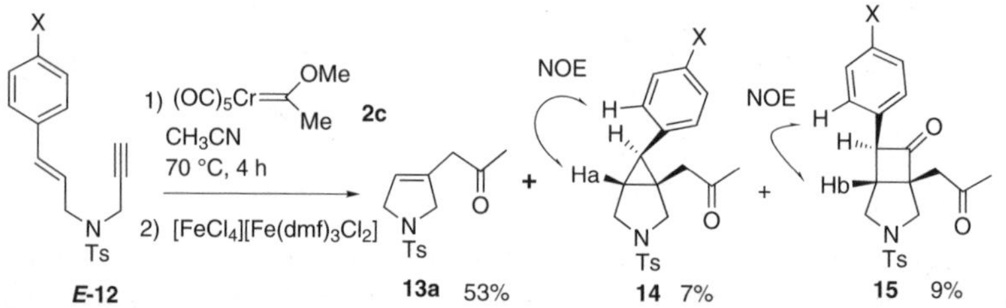

$$(4)$$

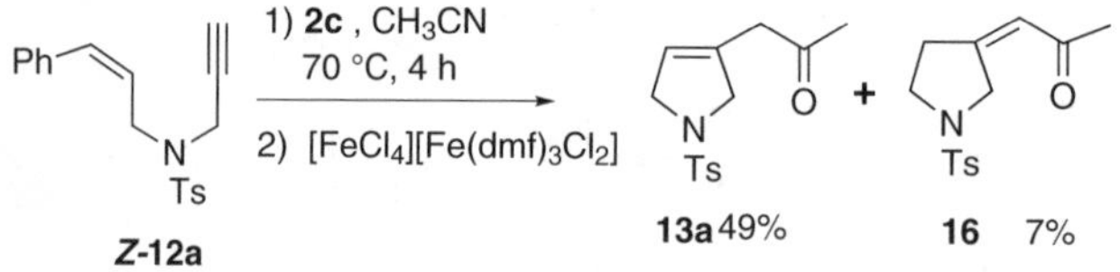

$$(5)$$

The reaction course is shown in Scheme 4. Enyne **12** reacts with **2** to give vinyl carbene complex **17**, which is in a state of equilibrium with vinyl ketene complex **21**. [2+2]Cycloaddition of the ketene moiety and alkene part in **21** gives cyclobutanone **22**. On the other hand, the vinyl carbene complex **17** reacts with the alkene intramolecularly to produce metalacyclobutane **18**. From metalacyclobutane **18**, reductive elimination occurs to give cyclopropane derivative **23**. Ret-

Table 2. Reaction of enyne 12 with Cr-carbene complex 2c

X	Substrate	13a(%)	14(%)	15(%)	13a:14
NO$_2$	12b	4	75	–	1:19
CI	12c	46	18	8	3:1
H	E-12a	53	7	9	7.7:1
Me	12d	62	–	6	1:0

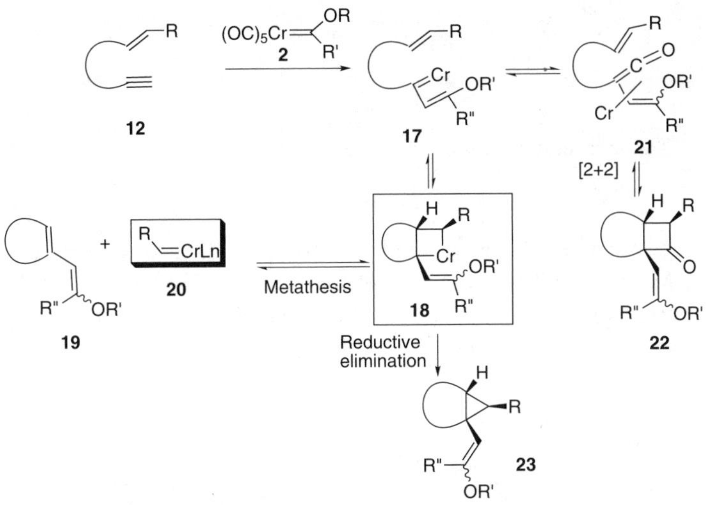

Scheme 4. Reaction mechanism of enyne with Fischer carbene complex

rocycloaddition occurs from 18 to give metathesis product 19 and alkylidene carbene complex 20. Since the [2+2]cycloaddition is controlled by HOMO-LUMO interactions, the cycloaddition in this instance should be retarded by a lack of activation of the double bond. Thus, 12b would afford only the chromacycle 18b, which gives the cyclopropane 23b (namely, 14b) in good yield because of the instability of the alkylidene carbene complex 20b. On the other hand, the electron donating groups on the double bond would favor the metathesis process because alkylidene carbene complex 20 generated from chromacycle 18 is stabilized by these substituents.

If the substituents on generated carbene complex 20 are the same as those on the alkene, this reaction must proceed by a catalytic amount of chromium carbene complex 2 (Scheme 5) [9].

To confirm this, enyne 12e was reacted with 30 mol% of chromium carbene complex 2c to give the metathesized products 24e, 13e, and 13a (Eq. 6). Although compound 13a was the reaction product of 12e and 2c, the former two products 24e and 13e were formed from enyne 12e and 2d, respectively. This indicates that chromium carbene complex 2d was generated in this reaction. On the basis of

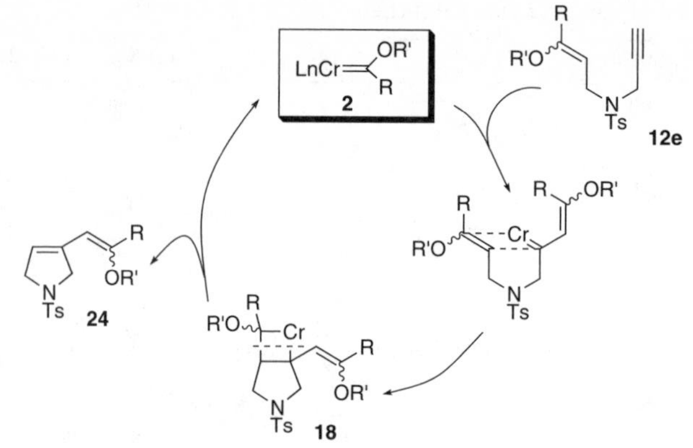

Scheme 5. Plan for chromium-catalyzed enyne metathesis

these results, enyne *E*-12f was reacted with 10 mol% of chromium carbene complex **2c** in MeOH to give metathesis product **13a** in 70% yield (Eq. 7). The reaction of *Z*-12f with 10 mol% of **2c** also gave **13a** in 45% yield (Eq. 8). In a similar manner, enyne **12g** gave **13g** in 25% yield (Eq. 9).

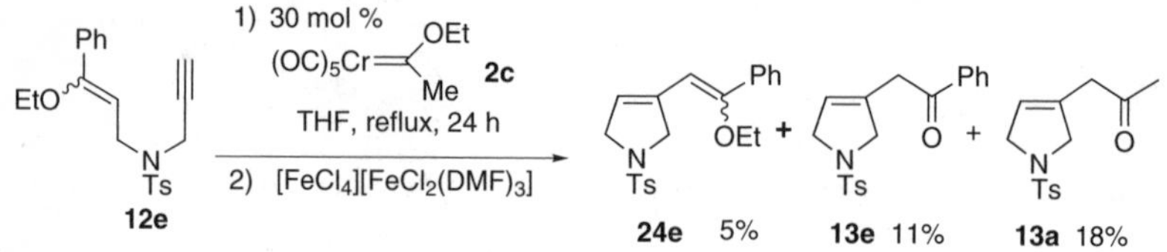

(6)

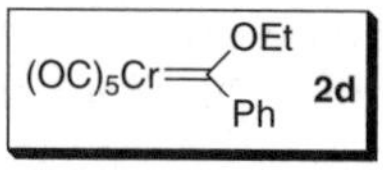

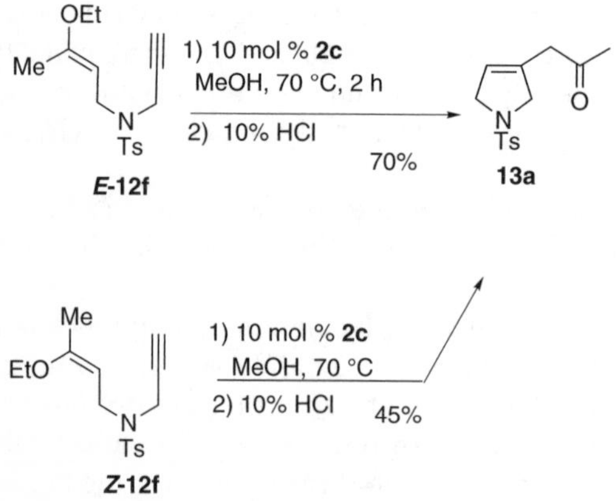

(7)

(8)

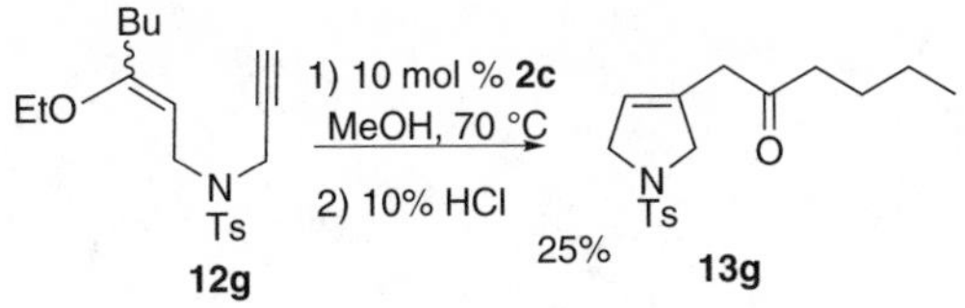

(9)

2.2
Pd- and Pt-Catalyzed Enyne Metathesis

Trost [10] discovered a palladium-catalyzed enyne metathesis during the course of his study on palladium-catalyzed enyne cyclization. Treatment of the 1,6-enyne **25** with palladacyclopentadiene (TCPT, **26a**) in the presence of tri-*o*-tolyl-phophite and dimethyl acetylene dicarboxylate (DMAD) in dichloroethane at 60°C led to cycloadduct **27** and vinylcyclopentene **28** in 97% yield in a ratio of 1 to 1 (Eq. 10). The latter compound **28** is clearly the metathesis product.

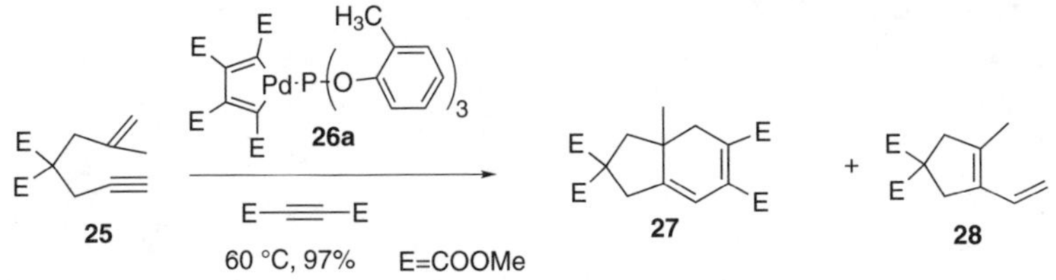

(10)

This reaction proceeds in a highly stereospecific manner.Thus, enyne *Z*-**29** gave a 68% yield of the diene product *E*-**31** (Eq. 11). Similarly, the E-substrate **29** gave predominantly *Z*-**31** (Eq. 12).

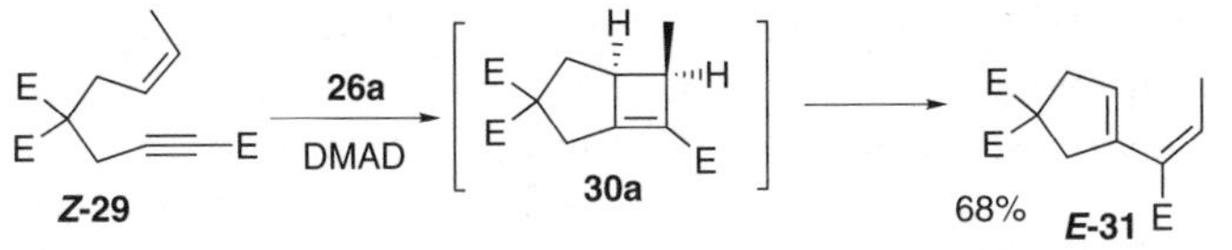

(11)

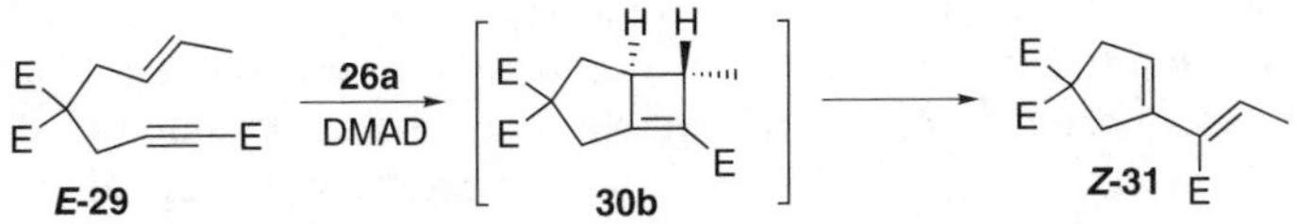

(12)

The reaction mechanism was considered to be oxidative cyclization, and palladacyclopentene **32** was formed. Reductive elimination then occurs to give cyclobutene **33**, whose bond isomerization occurs to give diene **28**. The insertion of alkyne (DMAD) into the carbon palladium bond of **32** followed by reductive elimination occurs to give [2+2+2]cocyclization product **27**. Although the results of the reactions of E- and Z-isomers of **29** with palladium catalyst **26a** were accommodated by this pathway, Trost considered the possibility of migration of substituents. Therefore, ^{13}C-labeled substrate **25-^{13}C** was used for this reaction.

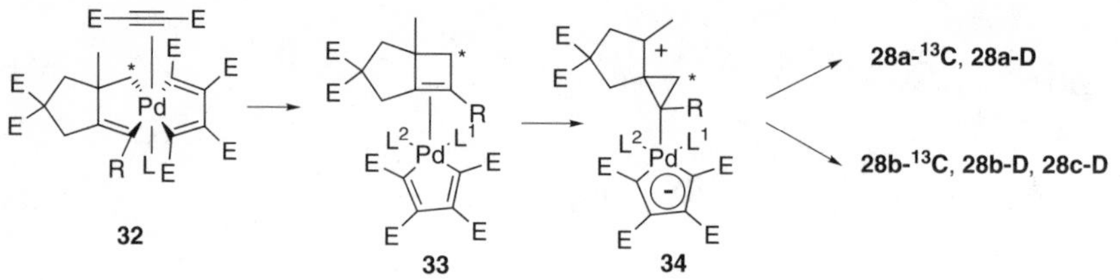

Scheme 6. Reaction mechanism

In the reaction of **25**-^{13}**C**, **28a**-^{13}**C** and **28b**-^{13}**C** were obtained in 52% yield along with 12% yield of **27**-^{13}**C** (Eq. 13). On the other hand, when **25-D** was treated in a similar manner, **28a-D**, **28b-D**, and **28c-D** were obtained in 42% yield with 19% yield of cycloadduct **27-D** (Eq. 14). From these results, one explanation for this reaction is shown in Scheme 6.

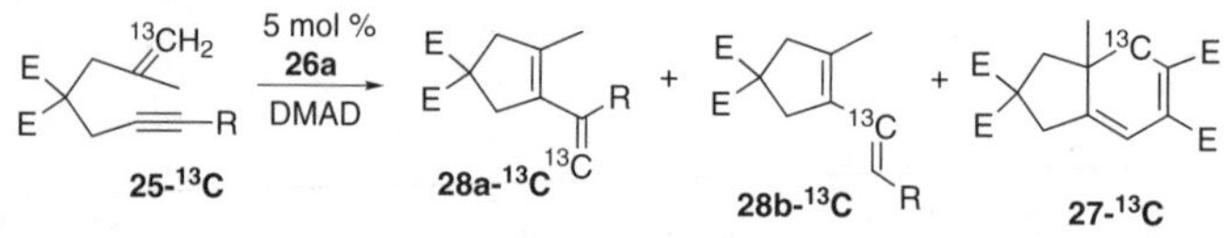

$$(13)$$

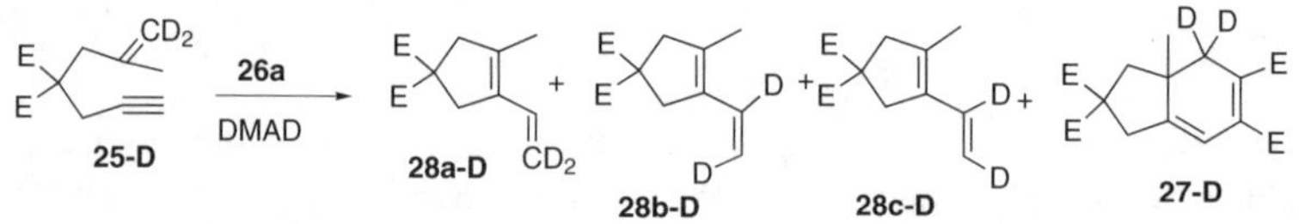

$$(14)$$

This method provides a very simple route to bridged bicycles possessing bridgehead olefins [11]. When enyne **35a** (n=1, 2, and 6) was treated with TCPC^{TFE} **26b**, seven-membered (**37a**, n=1, 53%), eight-membered (**37a**, n=2, 86%), and twelve-membered (**37a**, n=6, 73%) rings were formed (Eq. 15). Incorporation of nitrogen in the tether, as in **35b**, provides the azabicyclic compound **37b** in 58% yield [11a,b] (Eq. 16). In order to obtain further confirmation of the formation of a four-membered ring, a mixture of 4% TCPC^{HFB}, 4% o-tri-o-tolyl-phosphate, bis(heptafluorobutyl)-acetylenedicarboxylate, and enyne **35c** in dichloroethane was heated at 80°C, and cyclobutene **38** was obtained in 85% [11c] (Eq. 17). X-ray crystallography of **38** indicates that the stereochemistry of the fused 8,6-membered rings is *trans*. Although this palladium-catalyzed reaction seemed highly selective only in the case of substrates having an electron withdrawing group on the acetylenic terminal carbon and *cis* substituents on the olefic part, the reaction is very synthetically attractive.

$$(15)$$

TCPC^{TFE} (**26b**, R=CH$_2$CF$_3$)
TCPC^{HFB} (**26c**, R=CH$_2$CF$_2$CF$_2$CF$_3$)

$$(16)$$

$$(17)$$

A simple platinum complex also effects metathesis of enyoate as outlined in Eq. 18 [12]. The yield is comparable to that of TCPC but the reaction significantly faster. Murai and Chatani [13] also reported platinum-catalyzed conversion of 1,6- and 1,7-enyne to 1-vinylcycloalkenes. The treatment of 1,6-enyne **41a** with 4 mol% of PtCl$_2$ in toluene at 80°C under nitrogen for 3 h resulted in skeletal reorganization to give **42a** in 86% isolated yield (Eq. 19). The PtCl$_2$-catalyzed reaction of **44** (cyclic olefin) gave exclusively bicyclic compound **45** in 97% yield (Eq. 20). The reaction of **41b** with 8 mol% of PtCl$_2$ at 80°C was completed in 3 h to afford 86% of the two isomers **42b** and **43b** with a ratio of 8:1. The formation of **43** is interesting with respect to the reaction mechanism because it involves an unusual change in the bond relationship. While the olefinic terminal carbon in **41b** migrates onto the alkyne carbon bearing the methyl group to give **42b**, in a formal sense, insertion of the olefinic terminal carbon in **41b** into the C-C triple bond would give **43b**. The enyne **41c** having an ester group on the terminal alkene underwent skeletal reorganization to give exclusively **43c** in 80% yield. The anomalous C-C bond formation also occurred in the reaction of **41a-D**, and **42a-D** and **43a-D** were obtained in 86% yield in a ratio of 4 to 96 (Eq. 21).

$$(18)$$

$$(19)$$

41a R=H
41b R=Me
41c R=COOEt

42a, 42b   **43b, 43c**

$$(20)$$

44   **45**

$$(21)$$

41a-D   **42a-D**   **43a-D**

42a-D:*E*-**43a-D**:*Z*-**43a-D**=4:73:23

2.3
Ru-Catalyzed Enyne Metathesis

Murai and Chatani [14] also reported the ruthenium-catalyzed skeletal reorganization of 1,6-enynes **46** to 1-vinylcyclopentenes **47** (Eq. 22). They used $[RuCl_2(CO)_3]_2$ as a catalyst, and the reaction was carried out under an atmosphere of CO. $[RuBr_2(CO)_3]_2$, $RuCl_2(p\text{-cymene})]_2$ and $RuCl_3 \cdot xH_2O$ can be used for this reaction. The reactions of *E*-**46a** and *Z*-**46a** gave only the E-isomer of **47a**. The reaction of **46b** with $[RuCl_2(CO)_3]_2$ under CO gave *Z*-**47b** as a major product, although the reaction using Trost's catalyst gave a 1:1.4 mixture of **47b'** and bicyclo[3.2.0]heptene derivative, while **47b** was not formed [10] (Eq. 23). From these results, it must be concluded that this reaction differs from Trost's palladium-catalyzed reaction. This catalyst is also applicable to a 1,7-enyne **46c** (Eq. 24). The authors considered carbenoid **50** as an intermediate and attempted to trap this [15]. The reaction of 6,11-dien-1-yne **48** in toluene in the presence of 4 mol% of $[RuCl_2(CO)_3]_2$ at 80°C for 4 h under N_2 gave tetra-cyclic compound **49** in 84% yield (Eq. 25). The reaction involves two cyclopropanation reactions where both acetylenic termini formally act as a carbene.

$$(22)$$

46a   **47a**

E-**46a** (*E*/*Z*=80/20) 95%
Z-**46a** (*E*/*Z*=11/89) 81%

$$(23)$$

$$(Z\text{-}47b\text{:}E\text{-}47b\text{:}47b'=83\text{:}11\text{:}6)$$

$$(24)$$

$$(25)$$

Very reactive carbene complexes for olefin metathesis were recently discovered by Grubbs [2] and Schrock [3]. Grubbs et al. synthesized many cyclic compounds from dienes using his ruthenium carbene complex **52a** [5]. Using this ruthenium catalyst, he developed an elegant dienyne metathesis [16]. When a benzene solution of diene **51** and 3 mol% of **52a** was stirred at room temperature for 4 h, the cyclized product **53** was obtained in 90% yield (Eq. 26). In this reaction, dienyne **51** reacts with carbene complex **52a** to produce ruthenium carbene complex **54**, which reacts with olefin intramolecularly via [2+2] cycloaddition to produce ruthenacyclobutene **55** (Scheme 7). Retrocycloaddition occurs to give vinyl carbene complex **56**, which reacts again with the other olefin group intramolecularly to produce cyclized compound **53** and ruthenium methylene complex **57**, which is the real metathesis catalyst. It then reacts with dienyne **51** to regenerate carbene complex **54** and ethylene.

$$(26)$$

Cy=Cyclohexyl

The symmetrical dienyne **58a** was converted to a fused bicyclo[4.3.0] ring in 95% yield [17] (Eq. 27). With substrate **58c** containing an unsymmetrical diene tether, two different products, **59c** and **59c'**, were obtained in a ratio of 1 to 1 (Eq. 28). The reaction course in the formation of the different bicyclic rings is shown in Scheme 8. This dienyne metathesis is also catalyzed by tungsten or molybdenum complex **62** or **63** (Fig. 1), and a dienyne bearing terminal alkyne **58b** could be cyclized to give **59b** in 97% yield.

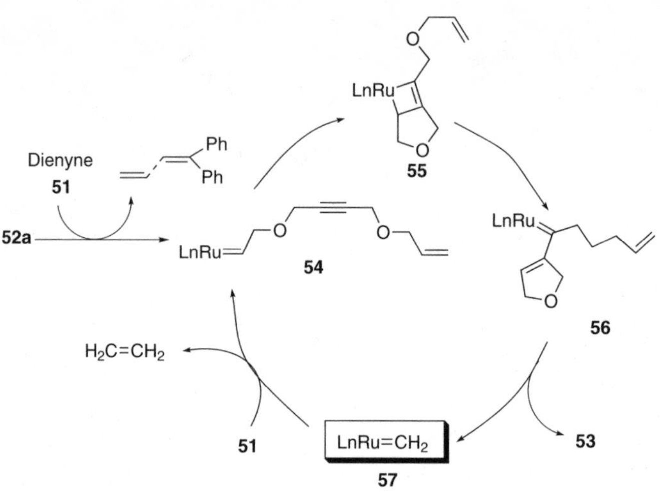

Scheme 7. Ruthenium-catalyzed dienyne metathesis

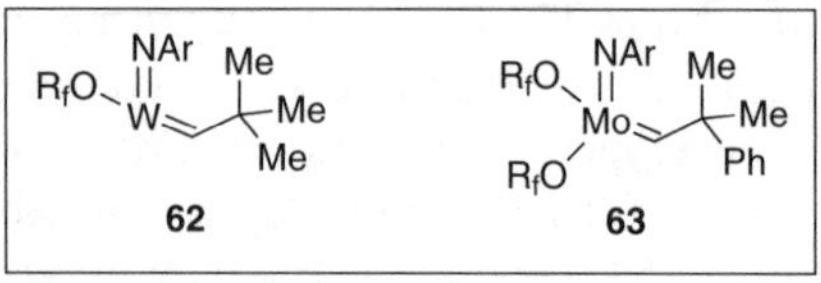

Fig. 1. Tungsten and molybdenum complexes **62** and **63**

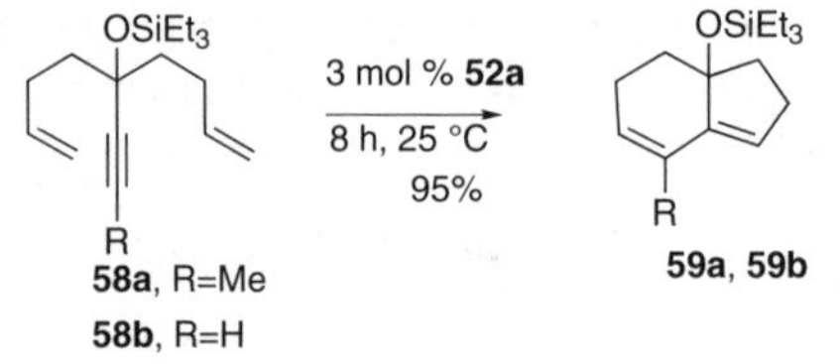

$$(27)$$

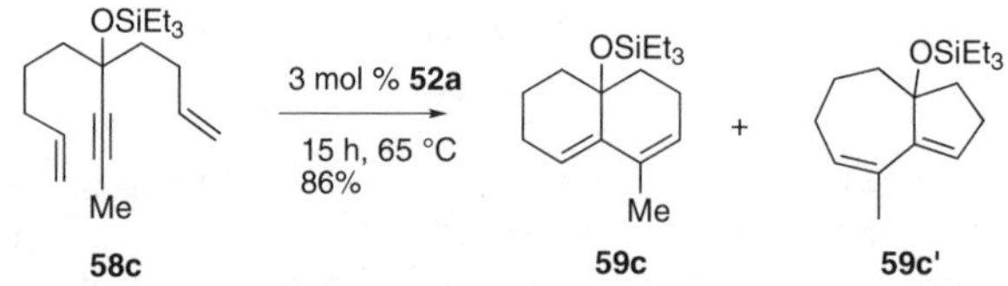

$$(28)$$

Enyne metathesis using ruthenium catalyst **52a** was developed by Mori and Kinoshita [18]. When enyne **62a** was treated with Grubbs's ruthenium catalyst **52a** in benzene at room temperature for 22 h, the cyclized product **63a** was obtained in only 13% yield (Eq. 29). It seems that the catalyst was coordinated by the diene generated in this reaction. This problem was overcome by the study of

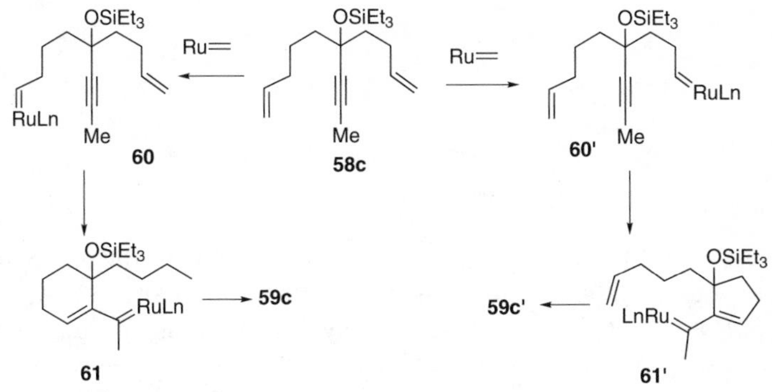

Scheme 8. Reaction mechanism

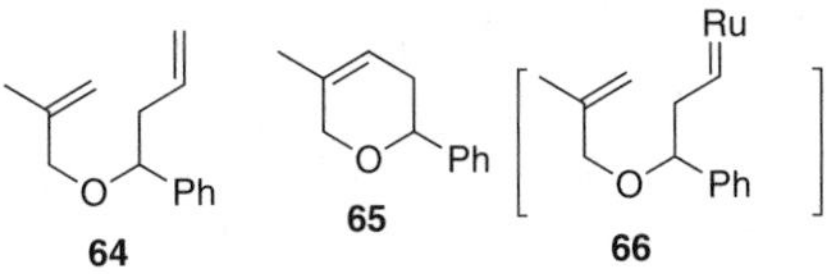

Fig. 2. Diene **64**, the cyclized product **65** and the alkylidene ruthenium complex **66**

the substituent effects on the olefin metathesis. When diene **64** was treated with 1 mol% of **52a**, the cyclized product **65** was not obtained. This means that alkylidene ruthenium complex **66** could not react with the exo-methylene part (Fig. 2). If the diene produced in enyne metathesis has an exo-methylene part, the ruthenium catalyst would not be coordinated by the diene moiety. On the basis of this idea, diene **63b** was obtained from enyne **62b** in 91% yield after only 35 min. Various cyclized products **63c–g** were obtained from the corresponding 1,6-, 1,7-, and 1,8-enynes **62c–g** having substituents on the alkyne (Table 3).

Table 3. Ruthenium-catalyzed enyne metathesis

Substrates			Time	Products	Yields (%)	
	62c	R=Ac	40 min		63c	86
	62d	R=TBDMS	50 min		63d	83
	62e	n=1	1.5 h		63e	88
	62f	n=1	1.5 h		63f	86
	62g	n=2	2.5 h		62g	77

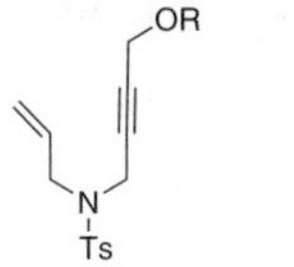

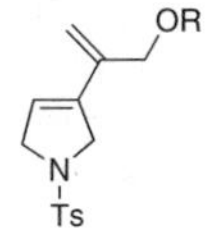

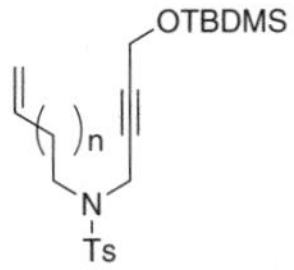

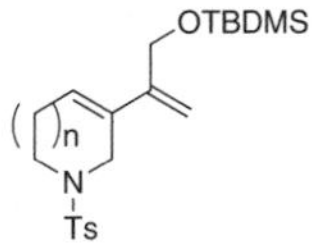

$$(29)$$

62a, R=H   **63a**, 13%
62b, R=Me   **63b**, 91%

To investigate if the ruthenium catalyst first reacts with the alkyne part or with the alkene part, the reaction of **64** with ruthenium catalyst **52a** was carried out (Eq. 30). Two metathesized products, enyne metathesis product **65** and diene metathesis product **66**, were obtained in 19% and 5% yields, respectively. This result indicates that the reaction of the alkyne part with **52a** is faster than that of the alkene part with **52a**.

$$(30)$$

64   **65** 19%   **66** 5%

3
Intermolecular Enyne Metathesis

Intermolecular-enyne metathesis, if it is possible, is very unique because the double bond of the alkene is cleaved and each alkylidene part is then introduced onto each alkyne carbon, respectively, as shown in Scheme 9. If metathesis is carried out between alkene and alkyne, many olefins, dienes and polymers would be produced, because intermolecular enyne metathesis includes alkene metathesis, alkyne metathesis and enyne metathesis. The reaction course for intermolecular enyne metathesis between a symmetrical alkyne and an unsymmetrical alkene is shown in Scheme 9. The reaction course is very complicated, and it seems impossible to develop this reaction in synthetic organic chemistry.

Compound **68** is known to form an electrically conducting "organic metal" with a large number of acceptor systems. When a toluene solution of enyne **67** and alkene **68** was refluxed for 2 days, the coupling product **70**, rather than the expected charge transfer complex, was obtained in good yield (Eq. 31, Table 4). This reaction involves a metathetic process [19].

$$(31)$$

67   **69**   **70**

68

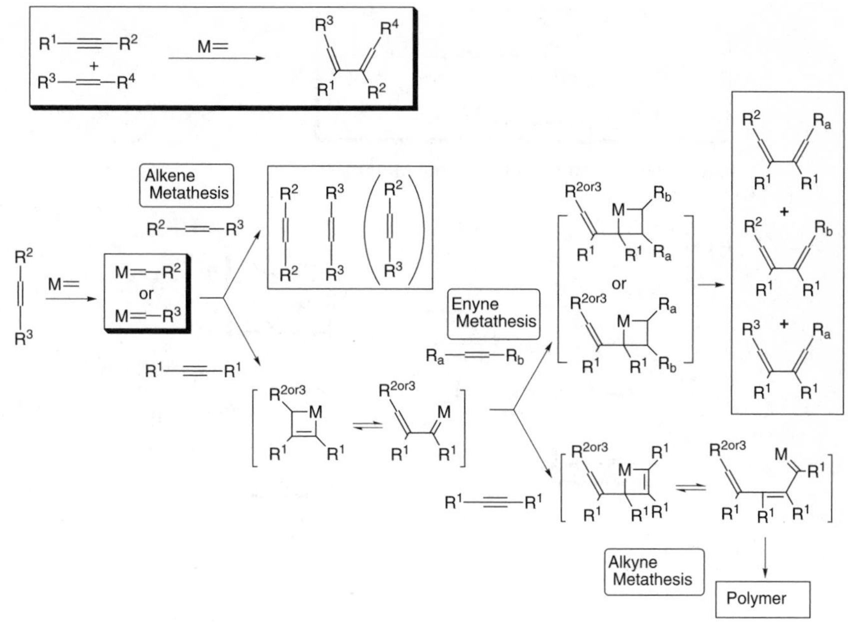

Scheme 9. Intermolecular enyne metathesis

Table 4. Reaction of **67** with **68**

R	Yield (%)	
CN	**70a**	77
Ph-C≡C	**70b**	77
COOEt	**70c**	47
H	**70d**	33

Intermolecular enyne metathesis has recently been developed using ethylene gas as the alkene [20]. The plan is shown in Scheme 10. In this reaction, benzylidene carbene complex **52b**, which is commercially available [16b], reacts with ethylene to give ruthenacyclobutane **73**. This then converts into methylene ruthenium complex **57**, which is the real catalyst in this reaction. It reacts with the alkyne intermolecularly to produce ruthenacyclobutene **74**, which is converted into vinyl ruthenium carbene complex **75**. It must react with ethylene, not with the alkyne, to produce ruthenacyclobutane **76** via [2+2]cycloaddition. Then it gives diene **72**, and methylene ruthenium complex **57** would be regenerated. If the methylene ruthenium complex **57** reacts with ethylene, ruthenacyclobutane **77** would be formed. However, this process is a so-called non-productive process, and it returns to ethylene and **57**. The reaction was carried out in CH_2Cl_2 un-

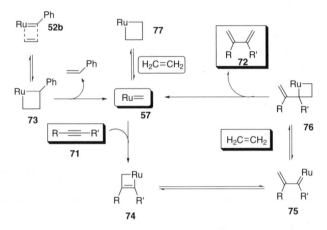

Pathway of Intermolecular Enyne Metathesis

Scheme 10. Plan for 1,3-diene synthesis using intermolecular enyne metathesis

der an atmosphere of ethylene gas at room temperature with a balloon, and the results are shown in Table 5. Symmetrical and unsymmetrical alkynes bearing functional groups in the tethers, and even the terminal alkynes, can be used in this reaction. In each case, the yield is moderate to good and the conversion yields are high. This procedure is a very useful and unique diene synthesis. Namely, the double bond of ethylene is cleaved and each methylene part is introduced into each alkyne carbon, respectively, as formally shown in Scheme 11.

Takahashi [21] reported a very interesting formal intermolecular enyne metathesis between alkyne and vinyl bromide using an equimolar amount of dibutyl zirconocene. When alkyne **71h** was treated with an equimolar amount of Cp_2ZrBu_2 in the presence of vinyl bromide, diene **72h** was obtained in 95% yield (Eq. 32). Cyclizations of **80** on zirconocene give diene **82** in high yields (Eq. 33). This reaction proceeds by the formation of zirconacycle **83** (Fig. 3) having a bromide at the α-position from alkyne **71h** and vinyl bromide, and bond isomerizations occur to give vinyl zirconium complex **84**, which is hydrolyzed to give diene **72h**.

$$Me_3Si\!\!=\!\!\!=\!\!\!-SiMe_3 \quad \xrightarrow[2. \;\; \diagup\!\!\!\!\diagdown_{Br}]{1. \; Cp_2ZrBu_2} \quad \overset{Me_3Si}{\diagdown}\!\!\!\!\overset{SiMe_3}{\diagup}$$

71h   95%   **72h**

(32)

Table 5. 1,3-Diene synthesis of intermolecular enyne metathesis

Run	Substrate	Ru (mol %)	Product	Yield[a]
1	 71a	3 10	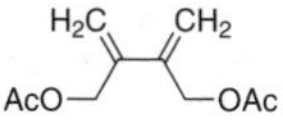 72a	66% (89%) 71% (90%)
2	 71b	3	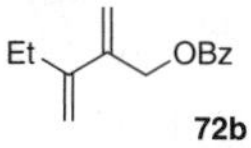 72b	62% (100%)
3	 71c	10[b]	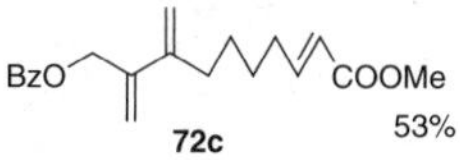 72c 53% (82%)	53% (82%)
4	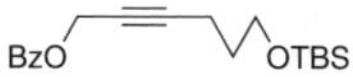 71d	10[b]	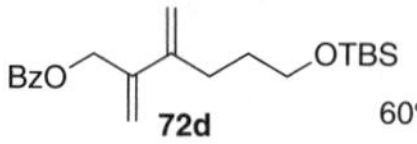 72d 60% (86%)	60% (86%)
5	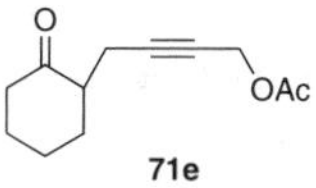 71e	3	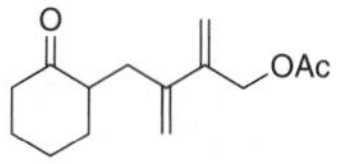 72e 74% (89%)	74% (89%)
6	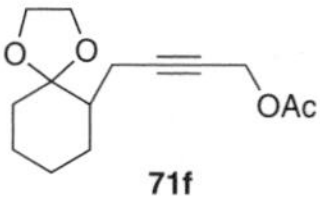 71f	3	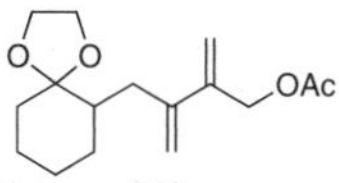 72f 48% (84%)	48% (84%)
7	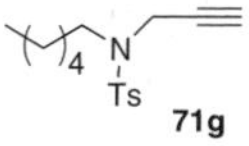 71g	5	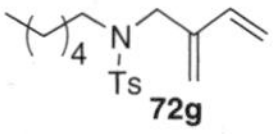 72g 81% (100%)	81% (100%)

All reactions were carried out in CH_2Cl_2 under ethylene gas using ruthenium catalyst **52b** at room temperature for 45 h.

[a] Yields in parentheses are conversion yields.
[b] 5 mol% of ruthenium complex catalyst **52b** was used. After 40 h, 5 mol% of **52b** was readded.

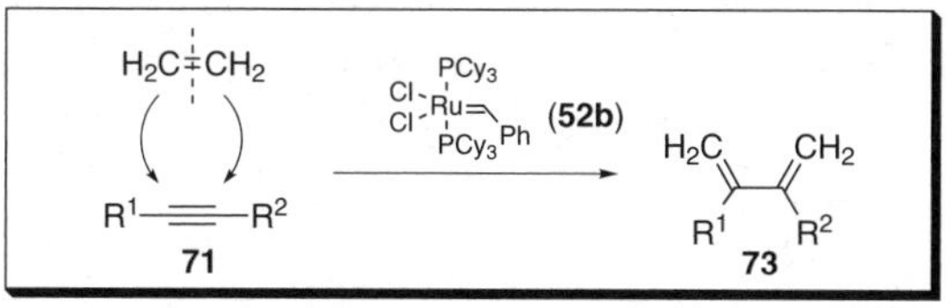

Scheme 11. 1,3-Diene synthesis from alkyne and ethylene

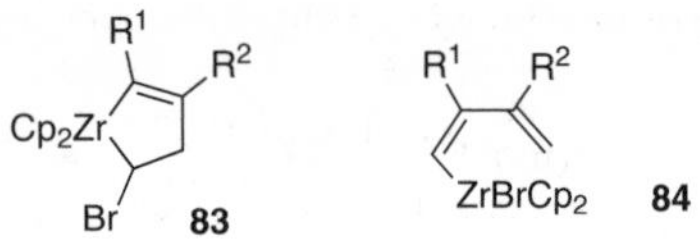

Fig. 3. Zirconacycle **83** and vinyl zirconium complex **84**

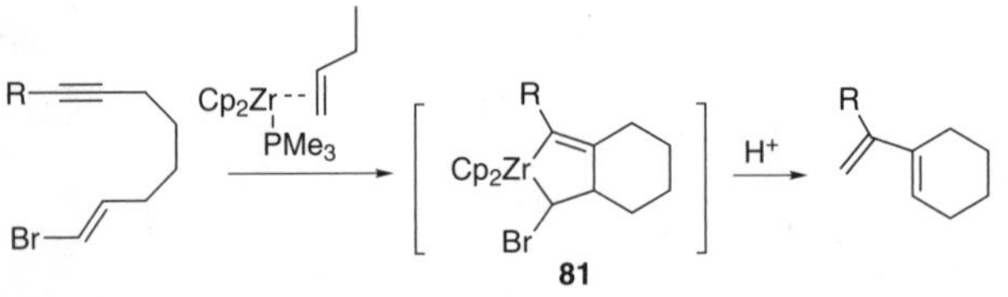

(33)

80a R=Ph
80b R=n-C$_8$H$_{17}$
80c R=SiEt$_3$

82a R=Ph, 84%
82b R=n-C$_8$H$_{17}$, 82%
82c R=SiEt$_3$, 87%

4
Utilization of Enyne Metathesis for the Synthesis of Natural Products and Biologically Active Substances

Since enyne metathesis has only recently been developed, there are only a few reports on the utilization of enyne metathesis in synthetic organic chemistry. Mori and Kinoshita [22] succeeded in the total synthesis of (-)-stemoamide (**85**) from (-)-pyrroglutamic acid using ruthenium-catalyzed enyne metathesis as a key step. The retrosynthetic analysis is shown in Scheme 12. The reaction of enyne **87a**, which was prepared from (-)-pyrroglutamic acid, with ruthenium carbene complex **52a** in benzene upon heating provided cyclized diene **86a** in 73% yield (Scheme 13). On the other hand, the reaction of enyne **87b** having a carbomethoxy group on the alkyne with **52b** in CH$_2$Cl$_2$ at room temperature gave diene **86b** in 87% yield. Reduction of **86b** followed by halolactonization gave bromide **90** and elimination product **91**. Treatment of **90** with NEt$_3$ gave **91**, which was reduced with NaBH$_4$ in MeOH in the presence of NiCl$_2$·6H$_2$O to give (-)-stemoamide (**85**).

Barrett [23] reported an ingenious method of synthesizing bicyclic β-lactams **93** using enyne metathesis (Scheme 14). From enynes **92** having various functional groups in the tethers, the desired bicyclic β-lactams **93** were obtained in good yields (Eqs. 34–36). In this reaction, terminal alkyne did not give a good result.

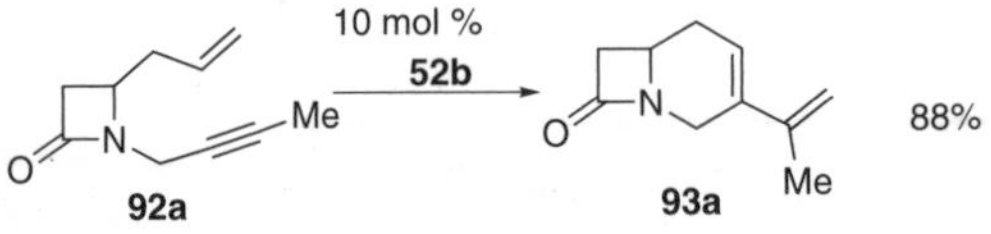

(34)

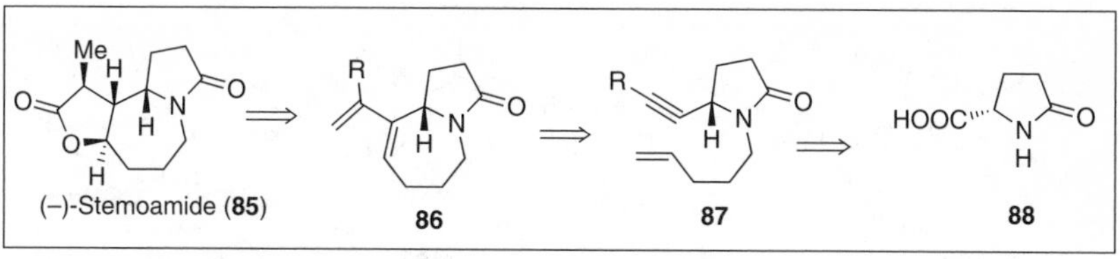

Scheme 12. Retrosynthetic analysis

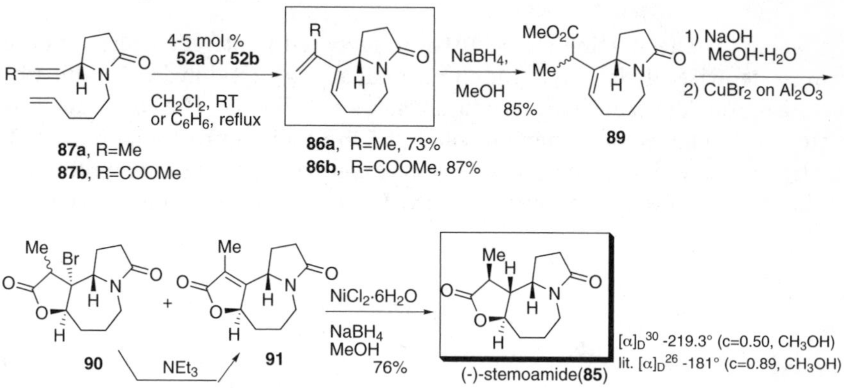

87a, R=Me
87b, R=COOMe

86a, R=Me, 73%
86b, R=COOMe, 87%

89

$[\alpha]_D^{30}$ -219.3° (c=0.50, CH$_3$OH)
lit. $[\alpha]_D^{26}$ -181° (c=0.89, CH$_3$OH)

Scheme 13. Synthesis of (-)-stemoamide

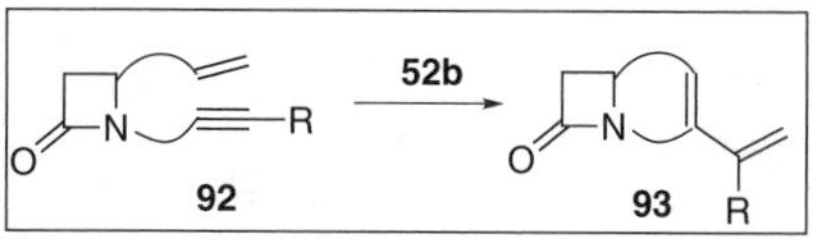

Scheme 14. Synthesis of bicyclic β-lactams

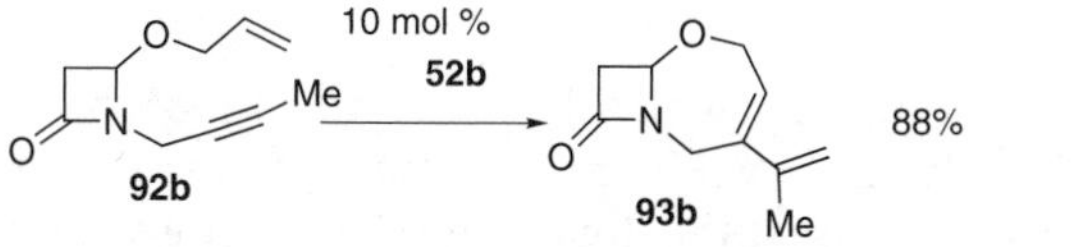

(35)

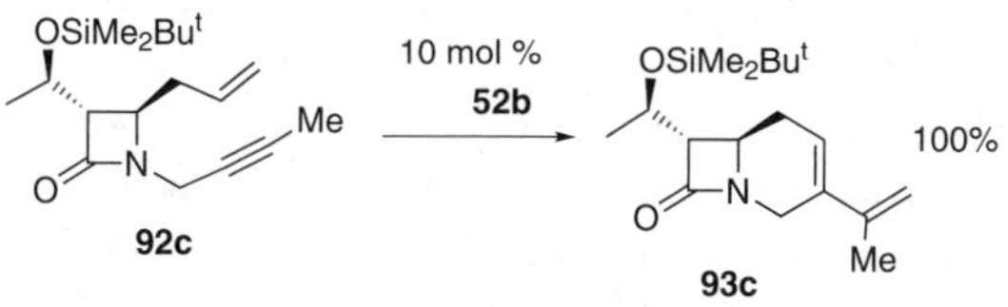

(36)

Undheim [24] described the stereoselective synthesis of cyclic 1-amino-1-carboxylic acid using ruthenium-catalyzed enyne metathesis. His plan is shown

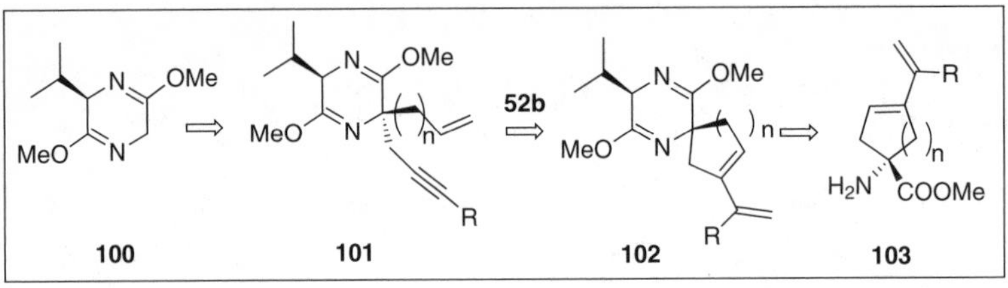

Scheme 15. Synthesis of cyclic 1-amino-1-carboxylic acid

in Scheme 15. The starting enyne **101** was prepared in stereochemically pure form by stepwise alkylations of the chiral auxiliary (R)-2,5-dihydro-3,6-dimethoxy-2-isopropylpyrazine with bromo-alkenes and -alkynes. The enyne **101** was treated with 5 mol% of ruthenium catalyst **52b** to give spiro compounds **102**, which were cleaved to the desired amino acid methyl ester **103** under mild acidic conditions (0.2 M trifluoroacetic acid in CH$_3$CN at room temperature) (Eqs 37, 38).

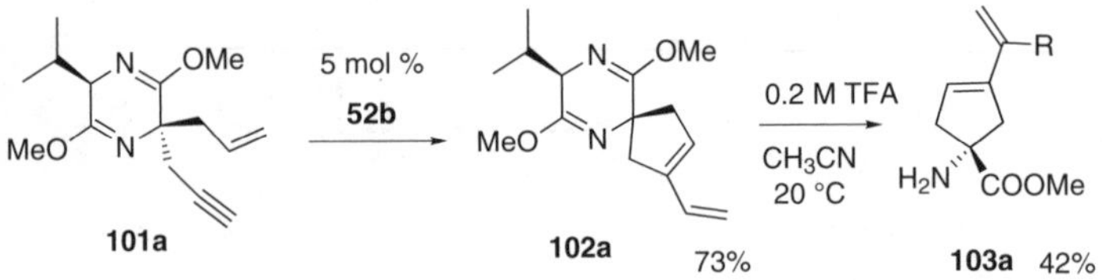

$$(37)$$

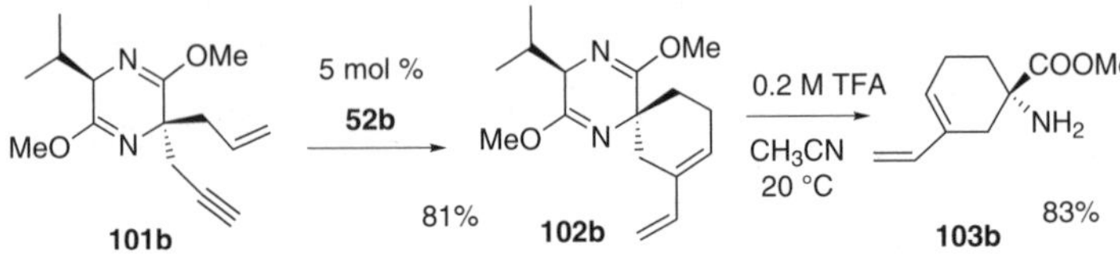

$$(38)$$

5
Perspective

Although this chemistry was first developed only 13 years ago, in 1985, by Katz, many interesting reactions have been shown in the literature. The main characteristic feature of enyne metathesis is that the double bond of alkene is cleaved and each alkylidene part is introduced onto respective alkyne carbons forming a compound with a diene moiety. Thus, the triple bond of alkyne is converted into a single bond. Among the many enyne metatheses or skeletal rearrangement reactions, ruthenium-catalyzed enyne metathesis is useful because the catalyst is commercially available, the reaction procedure is very simple, the reaction proceeds under very mild conditions, and the reaction mechanism is clear. The mechanisms of enyne metatheses using other metals are less obvious. Thus, many interesting reactions may be developed using various metal complexes.

Despite its difficulty, intermolecular enyne metathesis has been developed and will be a very important reaction in synthetic organic chemistry. It is expected that many interesting and useful applications will be developed in the future.

6
References

1. (a) Martin SF, Liao Y, Rein T (1994) Tetrahedron Lett 35:691. (b) Borer BC, Deerenberg S, Bieräugel H, Pandit UK (1994) Tetrahedron Lett 35:3191. (c) Martin SF, Wagman AS (1995) Tetrahedron Lett 36:1169. (d) Houri AF, Xu Z, Cogan D, Hoveyda AH (1995) J Am Chem Soc 117:2943. (e) Xu Z, Johannes CW, Salman SS, Hoveyda AH (1996) J Am Chem Soc 118:10926. (f) Fürstner A, Langemann K (1996) J Org Chem 61:8746. (g) Nicolaou KC, He Y, Vourloumis D, Vallberg H, Yang Z (1996) Angew Chem Int Ed Engl 35:2399. (h) Barrett AGM, Baugh SPD, Gibson VC, Giles MR, Marshall EL, Procopiou PA (1997) J Chem Soc Chem Commun 155. (i) Meng D, Su D-S, Balog A, Bertinato P, Sorensen EJ, Danishefsky SJ, Zheng Y-H, Chou T-C, He L, Horwitz SB (1997) J Am Chem Soc 119:2733. (j) Clark JS, Kettle JG (1997) Tetrahedron Lett 38:127. (k) Fürstner A, Kindler N (1996) Tetrahedron Lett 37:7005. (l) Fürstner A, Langemann K (1996) J Org Chem 61:3942. (m) Fürstner A, Langemann K (1997) Synthesis 792. (n) Fürstner A, Müller T (1997) Synlett 1010. (o) Chang S, Grubbs RH (1997) Tetrahedron Lett 38:4757. (p) Delgado M, Martin JD (1997) Tetrahedron Lett 38:6299. (q) Linderman RJ, Siedlecki J, O'Neill SA, Sun H (1997) J Am Chem Soc 119:6919. (r) Garro-Hélion F, Guibé F (1996) J Chem Soc Chem Commun 641
2. (a) Nguyen ST, Johnson LK, Grubbs RH, Ziller JW (1992) J Am Chem Soc 114:3974. (b) Nguyen ST, Grubbs RH, Ziller JW (1993) J Am Chem Soc 115:9858
3. Schrock RR, Murdzek JS, Bazan GC, Robbins J, DiMare M, O'Regan M (1990) J Am Chem Soc 112:3875
4. Recently, some elegant intermolecular olefin-metatheses were reported. (a) Crowe WE, Zhang ZJ (1993) J Am Chem Soc 115:10998. (b) Crowe WE, Goldberg DR, Zhang ZJ (1996) Tetrahedron Lett 37:2117. (c) Crowe WE, Goldberg DR (1995) J Am Chem Soc 117:5162. (d) Barrett AGM, Beall JC, Gibson VC, Giles MR, Walker GLP (1996) J Chem Soc Chem Comm 2229. (e) Barrett AGM, Baugh SPD, Gibson VC, Giles MR, Marshall EL, Procopiou PA (1994) J Chem Soc Chem Commun 2505. (f) Schuster M, Pemerstorfer J, Blechert S (1996) Angew Chem Int Ed Engl 35:1979. (g) Feng J, Schuster M, Blechert S (1997) Synlett 129. (h) Gibson SF, Gibson VC, Keen SP (1997) J Chem Soc Chem Commun 1107. (i) Snapper ML, Tallarico JA, Randall ML (1997) J Am Chem Soc 119:1478. (j) Tallarico JA, Bonitatebus Jr. PJ, Snapper ML (1997) J Am Chem Soc 119:7157
5. (a) Fu GC, Grubbs RH (1992) J Am Chem Soc 114:5426. (b) Fu GC, Grubbs RH (1992) J Am Chem Soc 114:7324. (c) Fu GC, Grubbs RH (1993) J Am Chem Soc 115:3800. (d) Fu GC, Nguyen ST, Grubbs RH (1993) J Am Chem Soc 115:9856. (e) Fujimura O, Fu GC, Grubbs RH (1994) J Org Chem 59:4029. (f) Grubbs RH, Miller SJ, Fu GC (1995) Acc Chem Res 28:446. (g) Miller SJ, Grubbs RH (1995) J Am Chem Soc 117:5855. (h) Leconte M, Jourdan I, Pagano S, Lefebvre F, Basset J-M (1995) J Chem Soc Chem Commun 857. (i) Shon Y-S, Lee TR (1997) Tetrahedron Lett 38:1283. (j) Armstrong SK, Christie BA (1996) Tetrahedron Lett 37:9373
6. (a) Katz TJ, Sivavec TM (1985) J Am Chem Soc 107:737. (b) Sivavec TM, Katz TJ (1989) Organometallics 8:1620
7. (a) Hoye TR, Suriano JA (1992) Organometallics 11:2044. (b) Korkowski PF, Hoye TR, Rydberg DB (1988) J Am Chem Soc 110:2676
8. (a) Mori M, Watanuki S (1992) J Chem Soc Chem Commun 1082. (b) Mori M, Watanuki S (1993) Heterocycles 35:679. (c) Watanuki S, Mori M (1995) Organometallics 14:5054

9. (a)Watanuki S, Ochifuji N, Mori M (1994) Organometallics 11:4129. (b) Watanuki S, Ochifuji N, Mori M (1995) Organometallics 14:5062
10. Trost BM, Tanoury GJ (1988) J Am Chem Soc 110:1636
11. (a) Trost BM, Trost MK (1991) J Am Chem Soc 113:1850. (b) Trost BM, Trost MK (1991) Tetrahedron Lett 32:3647. (c) Trost BM, Ynai M, Hoogsteen K (1993) J Am Chem Soc 115:5294. (d) Trost BM, Hashmi ASK (1994) J Am Chem Soc 116:2183. (e) Trost BM, Hashmi ASK (1993) Angew Chem Ed Engl 32:1085
12. Trost BM, Chang VK (1993) Synthesis 824
13. Chatani N, Furukawa N, Sakurai H, Murai S (1996) Organometallics 15:901
14. Chatani N, Morimoto T, Muto T, Murai S (1994) J Am Chem Soc 116:6049
15. Chatani N, Furukawa N, Seki Y, Kataoka K, Murai S (1997) Abstracts of 44th Symposium on Organometallic Chemistry, Osaka, p 370
16. (a) Grubbs RH, Miller SJ, Fu GC (1995) Acc Chem Res 28:446. (b) Schwab P, France MB, Ziller JW, Grubbs RH (1995) Angew Chem Int Ed Engl 34:2039. (c) Schwab P, Grubbs RH, Ziller JW (1996) J Am Chem Soc 118:100
17. (a) Kim SH, Bowden NB, Grubbs RH (1994) J Am Chem Soc 116:10801. (b) Kim SH, Zuercher WJ, Bowden NB, Grubbs RH (1996) J Org Chem 61:1073
18. Kinoshita A, Mori M (1994) Synlett 1020
19. Hopf H, Kreutzer M (1991) Angew Chem Int Ed Eng 30:1127
20. (a) Kinoshita A, Sakakibara N, Mori M (1997) Abstracts of 44th Symposium on Organometallic Chemistry p.326, Osaka. (b) Kinoshita A, Sakakibara N, Mori M (1997) J Am Chem Soc 119:12388
21. Takahashshi T, Xi Z, Fischer R, Huo S, Xi C, Nakajima K (1997) J Am Chem Soc 119:4561
22. (a) Kinoshita A, Mori M (1996) J Org Chem 61:8356. (b) Kinoshita A, Mori M (1997) Heterocycles (in press)
23. Barrett AGM, Baugh SPD, Braddock DC, Flack K, Gibson VC, Procopiou PA (1997) J Am Chem Soc 119:1375
24. Hammer K, Undheim K (1977) Tetrahedron 53:10603

Cross-Metathesis

Susan E. Gibson (née Thomas) and Stephen P. Keen

During the past 4 years the transition metal catalysed alkene cross-metathesis reaction has enjoyed increasing attention from organic chemists as a method for carbon–carbon bond formation. The impetus behind this was undoubtedly the development of well-defined functional group tolerant molybdenum and ruthenium alkylidene catalysts by Schrock and Grubbs respectively. In light of these recent advances, we review herein the cross-metathesis reactions of functionalised alkenes catalysed by well-defined metal carbene complexes. Acyclic cross-metathesis reactions are presented in a chronological order and are accompanied by discussion of mechanistic and selectivity issues. The formation of monomeric products from the related ring-opening cross-metathesis reaction is also covered and presented similarly.

Keywords: Cross-metathesis, Ring-opening, Alkenes, Catalysis, Dienes

1
Introduction

1.1
Scope

This review focuses on the cross-metathesis reactions of functionalised alkenes catalysed by well-defined metal carbene complexes. The cross- and self-metathesis reactions of unfunctionalised alkenes are of limited use to the synthetic organic chemist and therefore outside the scope of this review. Similarly, ill-defined multicomponent catalyst systems, which generally have very limited functional group tolerance, will only be included as a brief introduction to the subject area.

1.2
Ill-Defined Catalysts

Although the bulk of this review is concerned with well-defined metal carbene catalysts, it is important to note the contributions made to cross-metathesis chemistry by ill-defined or multicomponent catalysts. A brief discussion of the cross-metathesis reactions of functionalised alkenes using catalysts of this type will therefore be included here [1].

It was originally believed that extending the alkene metathesis reaction to functionalised substrates would pose a considerable problem since the classical catalyst system $WCl_6/EtAlCl_2/EtOH$ was intolerant of functional groups. A breakthrough came in 1972, however, when it was reported that the self-metathesis of methyl oleate 1 could be catalysed by a WCl_6/Me_4Sn catalyst [2]. Since this discovery, unsaturated esters, and in particular oleates, have been widely used as test substrates for determining the functional group tolerance of metathesis catalysts. For this reason, numerous catalysts have been used for the self-metath-esis of ω-unsaturated esters and there are many reports of cross-metathesis re-actions between esters of this type and simple alkenes [1]. An example, the cross-metathesis of methyl oleate 1 with ethene (ethenolysis) using a heteroge-neous rhenium catalyst [3], is shown below (Eq. 1).

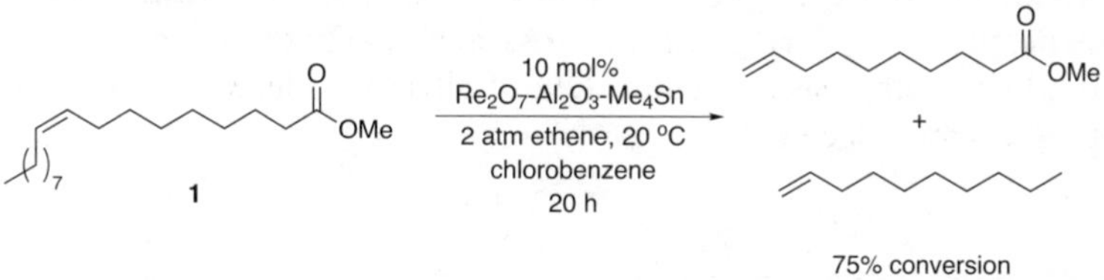

$$\tag{1}$$

75% conversion

Particularly noteworthy was the rhenium catalysed cross-metathesis of *trans*-hex-3-ene with vinyl acetate or α,β-unsaturated esters [4]. For example, cross-metathesis of methyl *trans*-crotonate with *trans*-hex-3-ene gave the desired cross-coupled product without any self-metathesis of the crotonate (Eq. 2).

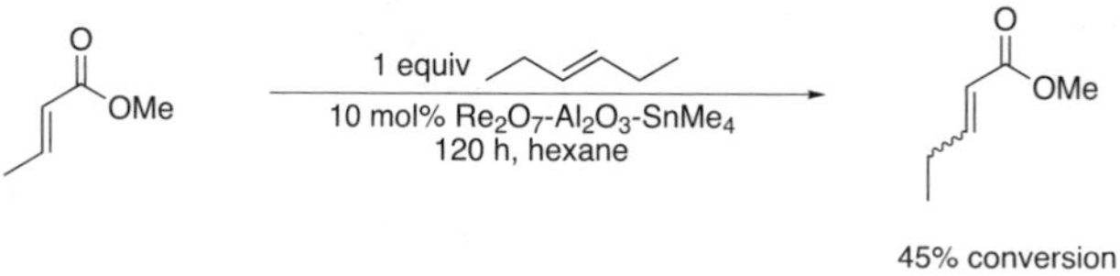

$$\tag{2}$$

45% conversion

The use of ill-defined catalysts for the cross-metathesis of allyl- and vinylsi-lanes has also received considerable attention, particularly within the past dec-ade. Using certain ruthenium catalysts, allylsilanes were found to isomerise to the corresponding propenylsilanes prior to metathesis [5]. Using rhenium- or tungsten-based catalysts, however, successful cross-metathesis of allylsilanes with a variety of simple alkenes was achieved [6, 7] (an example typical of the results reported is shown in Eq. 3).

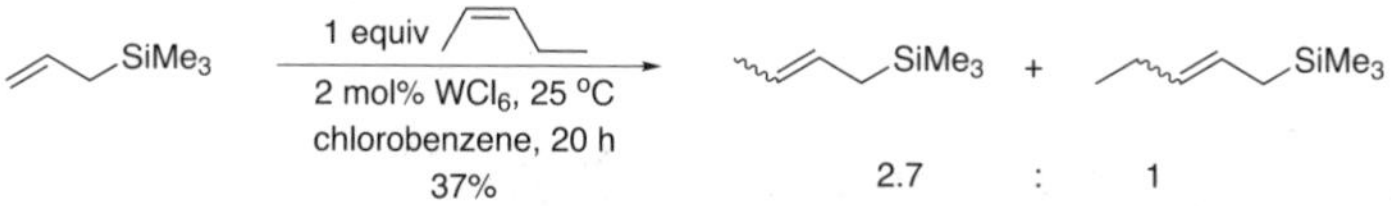

$$\tag{3}$$

In contrast, ruthenium catalysts gave the best results for the cross-metathesis reactions of vinylsilanes with a range of unfunctionalised alkenes [8] (a typical example is shown in Eq. 4).

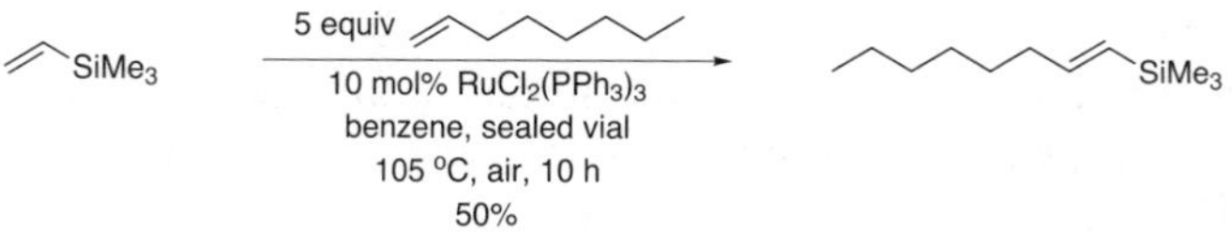

$$\tag{4}$$

Unsaturated esters and silanes are not the only functionalised alkenes to have been employed as cross-metathesis substrates: unsaturated alkyl chlorides [9], silylethers [10] and nitriles have all participated in metathesis reactions utilising

ill-defined catalysts. In fact, the cross-metathesis reactions of unsaturated nitriles have been reported several times [11] including, most recently, examples
of ring-opening cross-metathesis reactions [12]. Using a WCl_6/1,1,3,3-tetrame-
thyl-1,3-disilacyclobutane (DSCB) catalyst, cyclooctene and cyclopentene underwent ring-opening cross-metathesis with allyl cyanide to give the desired
monomeric dienes (for example 2). The formation of self-metathesis products
and dinitrile products, formed from cross-metathesis of the desired ring-
opened products with a second equivalent of allyl cyanide, was also observed
(for a typical example see Eq. 5).

$$(5)$$

2
Acyclic Cross-Metathesis

2.1
Alkylidene Catalysts

Initial reports of cross-metathesis reactions using well-defined catalysts were
limited to simple isolated examples: the metathesis of ethyl or methyl oleate with
dec-5-ene catalysed by tungsten alkylidenes [13, 14] and the cross-metathesis of
unsaturated ethers catalysed by a chromium carbene complex [15]. With the discovery of the well-defined molybdenum and ruthenium alkylidene catalysts 3
and 4, by Schrock [16] and Grubbs [17], respectively, the development of alkene
metathesis as a tool for organic synthesis began in earnest.

Ar = 2,6-diphenyl-C_6H_3

Prior to the first examples of the cross-metathesis of functionalised alkenes
using these catalysts, however, was a report on the use of a lesser known tungsten complex 5 [18, 19].

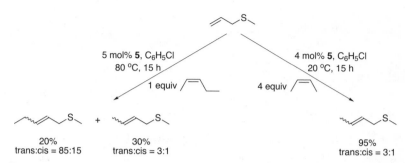

Scheme 1. Tungsten catalysed cross-metathesis of allyl methyl sulphide with pent-2-ene and but-2-ene

2.2
Tungsten Catalysed Cross-Metathesis

The successful cross-metathesis of allyl methyl sulphide with *cis*-pent-2-ene and *cis*-but-2-ene, catalysed by the tungsten alkylidene **5**, was reported by Basset and co-workers in 1993 [20] (Scheme 1).

Using an equimolar quantity of allyl methyl sulphide and *cis*-pent-2-ene resulted in incomplete reaction of the allyl sulphide and some self-metathesis of the sulphide substrate. When an excess (4 equiv) of but-2-ene was used, however, the desired but-2-enyl sulphide was formed in a good yield at ambient temperature. In this case, the large quantities of unwanted hydrocarbon starting material and self-metathesis products were gaseous alkenes and therefore easily removed. Using a large excess of one alkene to improve the yield of the desired cross-metathesis product in this way is obviously only viable if this alkene is inexpensive and both it and its self-metathesis product are easily removed.

Although the application of tungsten catalyst **5** to the cross-metathesis reaction of other alkenes has not been reported, Basset has demonstrated that ω-unsaturated esters [18] and glycosides [21], as well as allyl phosphines [22], are tolerated as self-metathesis substrates.

2.3
Selectivity with Styrene

Shortly after Grubbs' seminal communications on the application of molybdenum catalysed ring-closing metathesis to the synthesis of oxygen and nitrogen heterocycles [23, 24], the first cross-metathesis reactions using the Schrock catalyst **3** were reported [25]. Using just 1 mol% of the catalyst, styrene was rapidly cross-metathesised with several functionalised terminal alkenes to give products with >95% *trans* selectivity (for a typical example see Eq. 6).

$$\text{(6)}$$

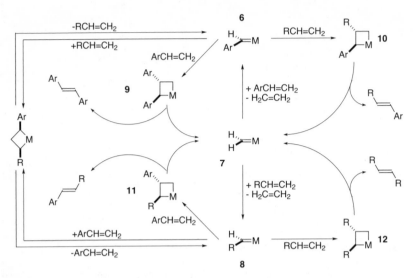

Scheme 2. Key intermediates for the four possible routes to the *trans* alkene products. Formation of the *cis* products occurs in the same manner, except via *cis* metallacyclobutanes. For the sake of clarity, the formation of non-productive metallacyclobutanes is not shown. For the same reason the reversible nature of all of these steps is also omitted

For the majority of substrates only trace amounts (<10%) of the self-metathesis products were isolated. Cross-/self-metathesis selectivity was significantly lowered, however, by the inductive effect of electron-withdrawing substituents on the alkyl-substituted alkene. Even moving a bromide one carbon closer to the double bond resulted in a significant decrease in the cross-/self-metathesis ratio (Eq. 7).

$$
\underset{\text{2 equiv } Ph}{\overset{\text{1 mol\% 3, DCM, 1 h}}{\longrightarrow}}
$$

$$
\begin{array}{cc}
n = 1 & 50\% \\
n = 2 & 90\%
\end{array}
\qquad
\begin{array}{c}
42\% \\
4\%
\end{array}
\tag{7}
$$

This correlation, between metathesis selectivity and the nucleophilicity of the alkyl-substituted alkene, was explained by considering the possible intermediate metallacyclobutanes **9–12** formed in the reaction (see Scheme 2).

Crowe proposed that benzylidene **6** would be stabilised, relative to alkylidene **8**, by conjugation of the α-aryl substituent with the electron-rich metal-carbon bond. Formation of metallacyclobutane **10**, rather than **9**, should then be favoured by the smaller size and greater nucleophilicity of an incoming alkyl-substituted alkene. Electron-deficient alkyl-substituents would stabilise the competing alkylidene **8**, leading to increased production of the self-metathesis product. The high *trans* selectivity observed was attributed to the greater stability of a *trans*-α,β-disubstituted metallacyclobutane intermediate.

The reaction tolerated a variety of functionality, including ester and ether groups on the alkyl-substituted alkene at least two carbons away from the double bond, and *meta*-nitro or *para*-methoxy substituents on the styrene. As expected, cross-metathesis occurred selectively at the less hindered monosubstituted double bond of dienes also containing a disubstituted alkene (Eq. 8).

$$\text{(8)} \qquad \frac{\text{2 equiv Ph} \diagdown}{\substack{\text{1 mol\% 3, DCM, 1 h} \\ 86\%}}$$

However, not all substrates readily underwent cross-metathesis. Alkenes containing ketone groups were found to be very poor substrates for the cross-metathesis reaction, although acceptable yields were obtained by protecting them as their silyl enol ethers prior to metathesis.

A year later, Schrock confirmed that the cross-metathesis of two alkyl-substituted terminal alkenes could also be catalysed by his molybdenum catalyst [26] (Eq. 9).

$$\text{(9)} \qquad \frac{\text{4.3 equiv hex-1-ene}}{\substack{\sim 1 \text{ mol\% 3, DME} \\ 3 \text{ h}}}$$

13
cis:trans = 1:4

14

13:14 = 11.5:1

The two alkenes were so similar electronically and sterically, with the ester group too far away to have any affect on the double bond, that there was very little cross-/self-metathesis selectivity. An approximately statistical mixture of ester **13** and diester **14** was isolated. The high yield of the cross-metathesis product **13** obtained is due to the excess of the volatile hex-1-ene used, rather than a good cross-/self-metathesis selectivity. Although not as predominant as in the reactions involving styrene, *trans* alkenes were still the major products.

2.4
cis Alkenes from Acrylonitrile Metathesis

In 1995 Crowe and co-workers underlined the potential of the molybdenum alkylidene **3** as a catalyst for cross-metathesis when they reported the first examples of productive acrylonitrile metathesis [27] (for example Eq. 10).

$$\text{(10)} \qquad \frac{\text{2 equiv NC} \diagdown}{\substack{\text{5 mol\% 3, DCM, 3 h} \\ 60\%}}$$

cis:trans = 7.6:1

Previously acrylonitrile had proved to be inert towards transition metal cata-lysed cross- and self-metathesis using ill-defined multicomponent catalysts [11b]. Using the molybdenum catalyst, however, acrylonitrile was successfully cross-metathesised with a range of alkyl-substituted alkenes in yields of 40–90% (with the exception of 4-bromobut-1-ene, which gave a yield of 17.5%). A dini-trile product formed from self-metathesis of the acrylonitrile was not observed in any of the reactions and significant formation (>10%) of self-metathesis products of the second alkene was only observed in a couple of reactions.

Like styrene, acrylonitrile is a non-nucleophilic alkene which can stabilise the electron-rich molybdenum-carbon bond and therefore the cross-/self-metathe-sis selectivity was similarly dependent on the nucleophilicity of the second alkene [metallacycle 10 versus 12, see Scheme 2 (replace Ar with CN)]. A notable difference between the styrene and acrylonitrile cross-metathesis reactions is the reversal in stereochemistry observed, with the *cis* isomer dominating (3:1–9:1) in the nitrile products. In general, the greater the steric bulk of the alkyl-substituted alkene, the higher the *trans/cis* ratio in the product (Eq. 11).

$$
\begin{array}{ll}
R = H & 79\% \quad \text{cis:trans} = 7:1 \\
R = Me & 64\% \quad \text{cis:trans} = 3:1
\end{array}
\tag{11}
$$

For a *cis* alkene to be formed the reaction would have to proceed through a *cis*-α,β-disubstituted metallacyclobutane intermediate (*cis* isomer of 10). Al-though it was unclear why there was a preference for forming a *cis* metallacycle, which leads to the thermodynamically less stable product, it was probably relat-ed to the small size or the electron-withdrawing properties of the nitrile group.

The success of the cross-metathesis reactions involving styrene and acryloni-trile led to an investigation into the reactivity of other π-substituted terminal alkenes [27]. Vinylboranes, enones, dienes, enynes and α,β-unsaturated esters were tested, but all of these substrates failed to undergo the desired cross-me-tathesis reaction using the molybdenum catalyst.

2.5
Allylsilanes: Good Cross-/Self-Metathesis Selectivity

Attacking the problem of cross-metathesis selectivity from a different angle, Crowe and co-workers explored the reactivity of a more nucleophilic partner for the π-substituted alkenes. They chose to use allyltrimethylsilane since they pro-posed that the CH_2SiMe_3 substituent should have a negligible effect on alkyli-dene stability, but enhance the nucleophilicity of the alkene via the silicon β-ef-fect (Fig. 1).

Cross-metathesis reactions with styrenes or acrylonitrile gave yields and *cis/trans* selectivities that were comparable with the best results obtained in the previous reports (for example Eq. 12).

Fig. 1. The silicon β-effect: stabilisation of a β-cation enhances the nucleophilicity of allyl-silanes

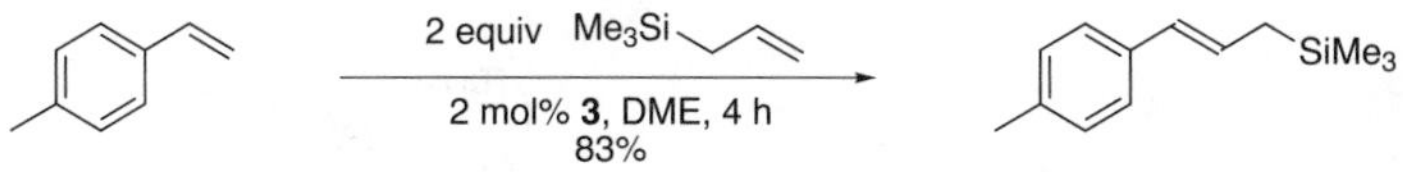

$$(12)$$

As expected, there was no formation of stilbenes or a dinitrile product and, more surprisingly, in all of the reactions reported only 5–7% of the allyltrimethylsilane self-metathesis product was observed. It was proposed that this lack of allylsilane self-metathesis was due to the steric bulk of the TMS group reducing the reactivity of the Me_3SiCH_2 substituted alkylidene. In a more recent report by Blechert and co-workers it was noted that allyltrimethylsilane and its hydrocarbon equivalent (4,4-dimethylpent-1-ene) had comparable reactivities in the cross-metathesis reaction [28], further suggesting that the selectivity arises from steric rather than electronic effects.

Cross-metathesis with small, functionalised, alkyl-substituted alkenes generally gave lower yields (34–73%) of the desired products, as predominantly their *trans* stereoisomers (2.5:1–5:1) [29] (for example Eq. 13).

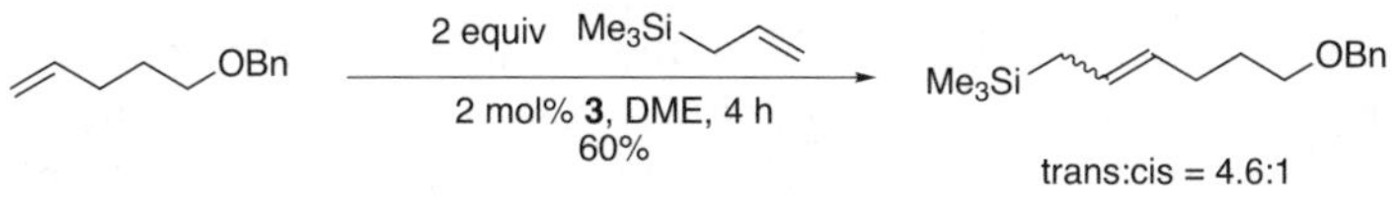

$$(13)$$

The ratio of cross-/self-metathesis products, with respect to the alkyl-substituted alkene, was generally poorer (typically 3:1) than the analogous reactions with styrene or acrylonitrile, probably due to the absence of a good alkylidene stabilising substituent on either alkene and the closer nucleophilicities of the two substrates.

From the single comparative study reported, it appears that increasing the steric bulk of the silyl group in the allyltrialkylsilane can significantly enhance the *trans* selectivity of the cross-metathesis reaction without adversely affecting the yield (Eq. 14).

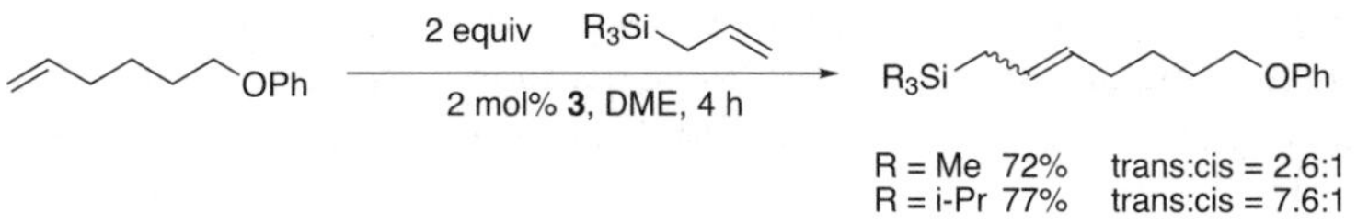

$$(14)$$

2.6
Homoallylic Alcohols and β-Lactams

During the past 2 years several research groups have published research that either uses or expands upon Crowe's acyclic cross-metathesis chemistry. The first reported application of this chemistry was in the synthesis of *trans*-disubstituted homoallylic alcohols [30]. Cross-metathesis of styrenes with homoallylic silyl ethers **15**, prepared via asymmetric allylboration and subsequent alcohol protection, gave the desired *trans* cross-metathesis products in moderate to good yields (Eq. 15).

$$R = Ph, Cy, Pr$$
$$Ar = Ph, p\text{-}ClC_6H_4, p\text{-}MeOC_6H_4$$

(15)

Barrett and Gibson also reported the application of the molybdenum catalysed cross-metathesis reaction to the elaboration of β-lactams [31, 32]. Protected allyloxy β-lactams **16** were successfully cross-metathesised with a selection of substituted styrenes to yield *trans* cross-metathesis products (Eq. 16).

$$R = Bn, TBDMS, CH(CO_2Et)OTBDMS$$
$$Ar = Ph, p\text{-}ClC_6H_4, p\text{-}MeOC_6H_4,$$
$$p\text{-}MeC_6H_4$$

(16)

It is worth noting that the reactions with the unsubstituted styrene gave noticeably higher yields than the corresponding cross-metathesis reactions with the substituted styrenes. This appears to be quite a common characteristic of the molybdenum catalysed cross-metathesis reaction.

2.7
Ruthenium Catalyst Mark II

Although the ruthenium vinylalkylidene catalyst **4** reported by Grubbs in 1993 had, in the years following its discovery, been widely used for ring-closing metathesis reactions, there were no reports on its use as a catalyst for the cross-metathesis of functionalised acyclic alkenes [33]. This was probably due to the generally lower metathesis activity of this catalyst **4** compared with Schrock's molybdenum catalyst **3**. The significantly lower turnover numbers observed with the ruthenium vinylalkylidene [17], compared with the molybdenum catalyst [26], for the self-metathesis of *cis*-pent-2-ene demonstrated this difference in activity. This was attributed in part to the slow initiation of the parent vinylalkylidene complex **4**.

With the development of an analogous ruthenium benzylidene catalyst **17** by Grubbs and co-workers in 1995, a ruthenium carbene catalyst suitable for the cross-metathesis reaction was in place [34]. Benzylidene **17** exhibited the same impressive tolerance of air and moisture, and the same stability towards functional groups as its predecessor **4**, but benefited from easier preparation [35, 36] and much improved initiation rates.

17

2.8
Cross-Metathesis with Polymer-Bound Substrates

The first published report on the use of this catalyst for the cross-metathesis of functionalised acyclic alkenes was by Blechert and co-workers towards the end of 1996 [37]. This report was also noteworthy for its use of polymer-bound alkenes in the cross-metathesis reaction. Tritylpolystyrene-bound N-Boc N-allylglycinol **18** was successfully cross-metathesised with both unfunctionalised alkenes and unsaturated esters (Eq. 17) (Table 1).

(17)

(P) = tritylpolystyrene; 0.2 mmolg⁻¹ loading

Two advantages of carrying out a cross-metathesis reaction with one of the alkenes immobilised on a polymer are clear from this study. Firstly, self-metathesis of the immobilised alkene can be suppressed, although this required (a) a low loading of the polymer (with 0.6 mmol g⁻¹ loading some self-metathesis was detected), and (b) use of a reasonably sterically bulky alkene (polymer-bound pent-4-enol or allyl alcohol readily self-metathesise). The second advantage is that a large excess of the soluble alkene can be used without causing difficulties in the isolation of the cross-metathesis product.

It is noticeable that cross-metathesis with the unfunctionalised alkenes occurred in significantly higher yields over shorter reaction times and required a smaller excess of the soluble alkene. This was possibly due to the unfunctionalised alkenes, which are more nucleophilic than their ester containing counterparts, complementing the less nucleophilic/more carbon-metal bond stabilising allylglycinol **18**. Comparable results were obtained from cross-metathesis reactions of the polymer-bound isomeric N-Boc C-allylglycinol with the same four alkenes.

In a second report by Blechert on the cross-metathesis of polymer-bound alkenes, an immobilised allylsilane **19** was reacted with a variety of highly func-

Table 1. Cross metathesis reactions of tritylpolystyrene-bound N-Boc N-allylglycinol

R	x equiv	Reaction time (h)	Conversion[b] (%)	trans:cis
Me[a]	3	12	>95	10:1
t-Bu	3	12	91	3:1
CO$_2$Me	6	24	62	6:1
OAc	6	24	65	10:1

[a] cis-hex-3-ene was used here as a less volatile propene equivalent.
[b] Determined by GC of the alcohol obtained after cleavage from the resin.

Scheme 3. Proposed mechanism for the protodesilylation of the polymer-bound metathesis products formed from allyl ethers or esters

tionalised alkenes in the presence of the Grubbs ruthenium benzylidene [38]. Subsequent cleavage of the metathesis products from the resin by protodesilylation with TFA gave one-carbon homologated products (for example Eq. 18).

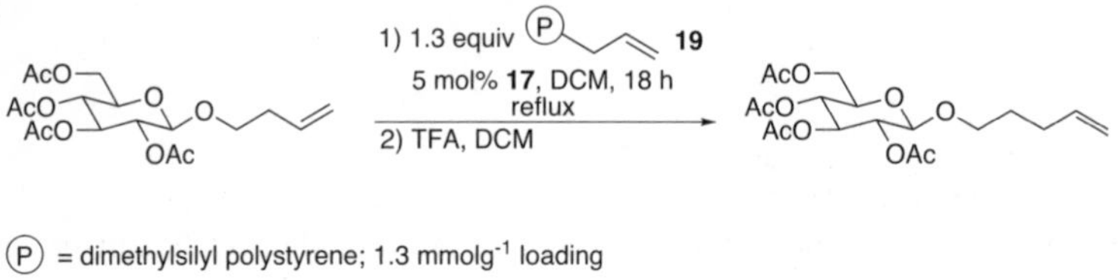

(18)

(P) = dimethylsilyl polystyrene; 1.3 mmolg^{-1} loading

Products of this type were not isolated, however, when the non-immobilised metathesis substrates were allyl esters or ethers. In these cases, the combined effect of metathesis and TFA cleavage was simply to remove the allyl group. A modified protodesilylation mechanism was proposed to account for these results (Scheme 3).

An example in which cleavage from the resin was achieved using a carbon electrophile demonstrated the ability of these polymer-bound metathesis products to act as substrates for carbon–carbon bond formation (Eq. 19).

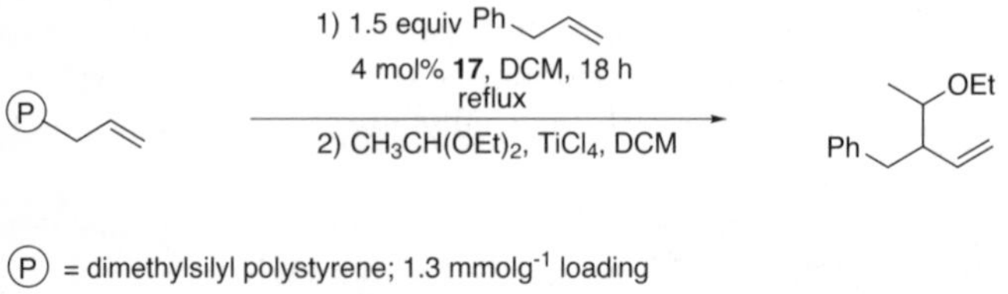

(19)

(P) = dimethylsilyl polystyrene; 1.3 mmolg^{-1} loading

2.9
Molybdenum Strikes Back

Although the Grubbs ruthenium benzylidene **17** has a significant advantage over the Schrock catalyst **3** in terms of its ease of use, the molybdenum alkylidene is still far superior for the cross-metathesis of certain substrates. Acrylonitrile is one example [28] and allyl stannanes were recently reported to be another. In the presence of the ruthenium catalyst, allyl stannanes were found to be unreactive. They were successfully cross-metathesised with a variety of alkenes, however, using the molybdenum catalyst [39] (for example Eq. 20).

$$\text{(20)}$$

trans:cis = 2.7:1

2 equiv Ph$_3$Sn

5 mol% **3**, DCM, 12 h
reflux
78%

Although these reactions required higher temperatures, longer reaction times (up to 48 h) and in some cases greater quantities of catalyst (10 mol%) than the previously reported molybdenum catalysed cross-metathesis reactions, moderate to good yields were obtained with a number of highly functionalised alkenes. Successful metathesis was achieved with alkenes containing ester, malonate, nitrile, acetal, glycoside or carbamate functional groups, which suggests that the molybdenum catalyst is not as sensitive to functionalised substrates as it was first believed. Two alkenes that failed to give any cross-metathesis products in these reactions, N-benzyl-4-vinyloxazolidin-2-one and allyl isocyanate, were, however, noted to be reactive substrates for the cross-metathesis with allyltrimethylsilane using the molybdenum and ruthenium catalysts respectively.

As with the allylsilane cross-metathesis reactions, significant quantities of allyl stannane self-metathesis were not detected in any of the reactions and the *trans* isomer predominated in the cross-metathesis products. Identical reactions were carried out using allyltributyl stannane, in place of allyltriphenyl stannane, but the yields of the cross-metathesis products were consistently lower and in many cases dropped below 25%.

2.10
Good Metathesis Selectivity from Sterically Bulky Substrates

Blechert and co-workers have also reported the application of cross-metathesis to the synthesis of jasmonic acid derivatives containing modified alkene side chains [28]. Molybdenum or ruthenium catalysed cross-metathesis of acetal **20** with various alkenes gave the desired cross-metathesis products in high yields (Eq. 21) (Table 2).

Table 2. Cross metathesis reactions of acetal **20**

R	x equiv	Catalyst[a]	Reaction time (h)	Yield (%)	trans:cis
CH_2CO_2Me	1.5	5 mol% Mo	5	83	2:1
CH_2OAc	2	5 mol% Ru	20	73	2:1
CH_2CH_2OH	1.5	8 mol% Ru	8	70	6:1
CN	2	5 mol% Mo	2	92	0:1
CH_2t-Bu	1.5	5 mol% Mo	3	65	4:1

[a] Mo, molybdenum alkylidene **3**; Ru, ruthenium benzylidene **17**.

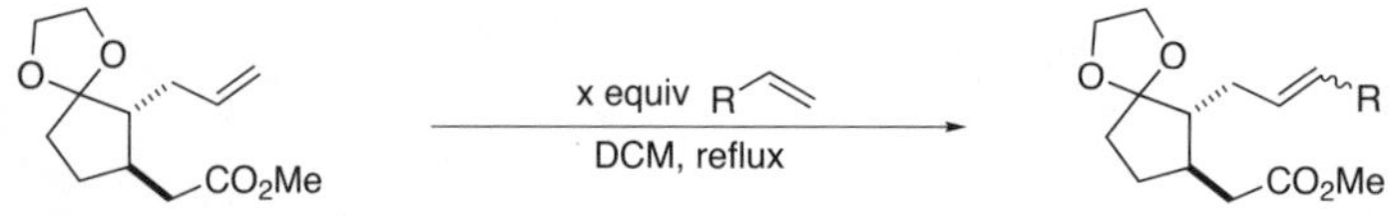

$$(21)$$

20

Performing the cross-metathesis reactions at reflux (open to an argon atmosphere in a glove box) was designed to help remove the unwanted ethene produced by the reaction.

No self-metathesis products of the acetal **20** were isolated from these reactions, probably due to the large steric bulk of this substrate preventing formation of an α,β-disubstituted metallacyclobutane containing two of these substituents. Using a sterically bulky alkene appears to be another way of ensuring good cross-/self-metathesis selectivity, provided it is not so large as to prevent it from reacting at all. Self-metathesis of the second alkene ($RCH=CH_2$) was observed to some extent with methyl but-3-enoate (26%) and allyl acetate (12%), but only small amounts, if any, were observed with the other three alkenes.

It is interesting to note that the two reactions involving allyl acetate and the unprotected alcohol, but-3-en-1-ol, failed when the molybdenum catalyst was used. The failure of the Schrock catalyst to tolerate unprotected alcohols has also been observed in ring-closing metathesis [40], where a tertiary alcohol has proved to be the only success [41].

Using the ketone instead of the acetal **20** caused a considerable drop in the yield of the cross-metathesis products, which was due in some cases to competing olefination of the ketone (Scheme 4).

Some further examples of the allyltrimethylsilane cross-metathesis reactions, originally investigated by Crowe, with more complex substrates were also in-

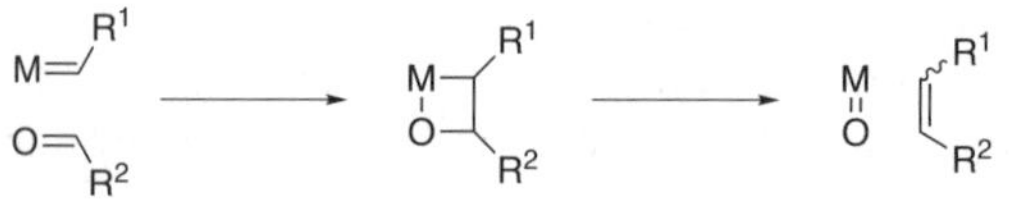

Scheme 4. Ketone olefination: reaction of a ketone with a metal alkylidene, which results in the destruction of the catalytic species

cluded in this paper [28]. Two particularly notable examples are given below (Eqs. 22, 23). The first demonstrates that, with the less active ruthenium benzylidene catalyst, it is possible to regioselectively cross-metathesise allyltrimethylsilane with the less hindered of two monosubstituted double bonds.

$$\text{(22)}$$

trans:cis = 1.5:1

2.11
Metathesis and α-Amino Acids

The second example was the startlingly successful cross-metathesis of allyltrimethylsilane with a sterically hindered, readily isomerised vinylglycine **21** in an excellent yield and with very little isomerisation using Schrock's molybdenum alkylidene (Eq. 23).

$$\text{(23)}$$

21, 97% ee   92% ee

About the same time, we published our own results on the cross-metathesis of the amino acid homoallylglycine using the Grubbs ruthenium catalyst **17** [42]. Both styrene and oct-1-ene were successfully cross-metathesised with protected homoallylglycine to give the desired products in moderate to good yields (Eq. 24).

$$\text{(24)}$$

R^1 = Boc, Phth, Ac, Fmoc
R^2 = Me, Bn, t-Bu, H

R^3 = Ph; 43–55%
          hexyl; 58–66%

The cross-metathesis reaction was successfully carried out using a variety of protecting groups including Boc, Fmoc, acetyl and phthaloyl amine protecting groups and benzyl and *t*-butyl esters. Even an unprotected carboxylic acid was tolerated by the catalyst. In all cases self-metathesis of the homoallylglycine made up the remainder of the yield and only a trace of any unreacted amino acid was observed. The ruthenium catalysed cross-metathesis reactions with styrene, in analogy with the molybdenum catalysed reactions reported by Crowe, yielded

Table 3. A comparison of the cross metathesis reactions between oct-1-ene and homoallyl-, allyl- and vinylglycine

	n	Isolated yields (%)		
		Cross-metathesis	Amino acid self-metathesis	Starting material
22	2	66	28	0
23	1	45	16	31
24	0	7	0	69

cross-metathesis products with very high *trans/cis* ratios and afforded only small amounts of stilbene.

The reactions were performed under a steady stream of nitrogen to aid the removal of ethene from the reaction mixture. Changing from a static nitrogen atmosphere to this flow of nitrogen resulted in a 30% yield enhancement and appeared to extend the life of the catalyst.

We have also studied the effect that moving the double bond closer to the amino acid moiety has upon the reactivity of unsaturated α-amino acids [43]. To this end, the cross-metathesis reactions of similarly protected homoallyl-, allyl- and vinylglycine with oct-1-ene were investigated under identical conditions (Eq. 25) (Table 3).

$$\tag{25}$$

It appears that the molybdenum catalyst is more suited to the cross-metathesis of the sterically bulky vinylglycines. The cross-metathesis reaction of a similarly protected dehydroalanine gave only recovered starting material.

Although the metathesis reaction with allylglycine **23** did not go to completion, a moderate yield of the desired cross-metathesis product was isolated. Very recently, Blechert has reported two similar cross-metathesis reactions of an allylglycine **25** using the ruthenium catalyst [44]. In these cases higher yields of the cross-metathesis products were isolated, presumably due to the higher reaction temperatures employed (Eq. 26).

$$\tag{26}$$

n = 1 or 8

2.12
Combinatorial Synthesis via Cross-Metathesis

Applications of the cross-metathesis reaction in more diverse areas of organic chemistry are beginning to appear in the literature. For example, the use of alkene metathesis in solution-phase combinatorial synthesis was recently reported by Boger and co-workers [45]. They assembled a chemical library of 600 compounds 27 (including *cis/trans* isomers) in which the final reaction was the metathesis of a mixture of 24 ω-alkene carboxamides 26 (prepared from six aminodiacetamides, with differing amide groups, each functionalised with four ω-alkene carboxylic acids) (Eq. 27).

$$(27)$$

26

n = 3, 4, 7 and 8

27

A complete set of sublibraries, 15 containing 72 compounds each and 6 containing 20 compounds each, were prepared in a similar manner, with comparable yields (37–75%).

2.13
Functionalisation of Si/O Frameworks

Alkene cross-metathesis has also been recently used for the modification of silsesquioxanes and spherosilicates, by Feher and co-workers [46]. Reaction of vinylsilsesquioxane 28 with a variety of simple functionalised alkenes, in the presence of Schrock's molybdenum catalyst 3, gave complete conversion of the starting material and very good isolated yields of the desired products (75–100%) (for example Eq. 28).

$$(28)$$

28

†per SiCH=CH$_2$ group

The relatively long reaction times employed and, for some reactions, the large excess of the second alkene required are possibly due to the sheer size of the silsesquioxane framework hindering reactivity at the vinylsilane sites. Presumably the absence of vinylsilsesquioxane self-metathesis products in all of the reactions is also due to the steric demands of the Si/O framework. Similar results were observed for the cross-metathesis of the analogous vinylspherosilicate $[(H_2C=CHSiMe_2O)_8Si_8O_{12}]$ with either styrene or pent-1-ene. Using the less active ruthenium benzylidene **17**, cross-metathesis occurred very slowly and poor conversions ($\leq 25\%$) of vinylsilsesquioxane into the desired cross-coupled products were observed.

It is worth noting that the reactions were run under a slight vacuum to aid the removal of the ethene produced during the reaction, which offers an alternative to performing the reaction at reflux or under a stream of nitrogen.

3
Ring-Opening Cross-Metathesis

3.1
What Is Ring-Opening Cross-Metathesis?

The ring-opening cross-metathesis reaction is similar to the acyclic cross-metathesis reaction discussed above, except that one of the acyclic alkenes is replaced with a strained cyclic alkene (Scheme 5).

Since one of the substrates is a cyclic alkene there is now the possibility of ring-opening metathesis polymerisation (ROMP) occurring which would result in the formation of polymeric products **34** (n >1). Since polymer synthesis is outside the scope of this review, only alkene cross-metathesis reactions resulting in the formation of monomeric cross-coupled products (for example **30**) will be discussed here.

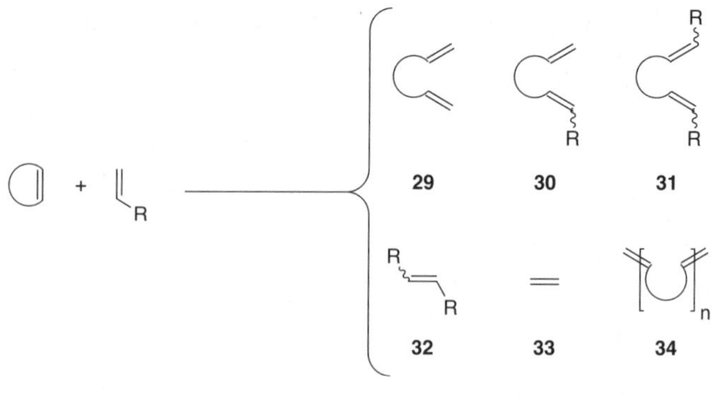

Scheme 5. Possible metathesis products from the reaction of a cyclic and an acyclic alkene

3.2
Tungsten Catalysed Ring-Opening Metathesis

The report by Basset and co-workers on the metathesis of sulphur-containing alkenes using a tungsten alkylidene complex, mentioned previously for the acyclic cross-metathesis reaction (see Sect. 2.2), also contained early examples of ring-opening cross-metathesis of functionalised alkenes [20]. Allyl methyl sulphide was reacted with norbornene in the presence of the tungsten catalyst 5, to yield the desired ring-opened diene 35 (Eq. 29).

$$\text{(29)}$$

Unfortunately, this product was isolated as a mixture with diene 36, formed from cross-metathesis with a second equivalent of the allyl sulphide, and was contaminated with some polymeric residues. It is also important to note that an excess of the sulphide was required to suppress competing ROMP of the norbornene. A similar result was obtained for the reaction of allyl methyl sulphide with cyclopentene.

The reaction also tolerated the thiol group on the cyclic substrate: butylthiocyclooctene 37 was ring-opened in the presence of an excess of ethene to give a good yield of the diene 38 (Eq. 30).

$$\text{(30)}$$

In this case, the use of ethene as the acyclic alkene means that the diene 38 and polymeric compounds are the only possible products that can be formed from metathesis.

3.3
Ring-Opening Cross-Metathesis of Cyclobutenes

3.3.1
The Reactions

In 1995 the first examples of ring-opening cross-metathesis reactions for the preparation of functionalised monomeric products using the Grubbs ruthenium vinylalkylidene catalyst 4 were published by Snapper and co-workers [47]. Reaction of a variety of symmetrical cyclobutenes with simple terminal alkenes yielded the desired cross-metathesis products in good yields (for example Eq. 31).

$$\text{cis:trans} = 2.3:1$$

(31)

Optimal yields were obtained by slow addition of the alkene substrates to a solution of the ruthenium vinylalkylidene and this allowed just two equivalents of the acyclic alkene to be used without significant formation of polymeric products. Unlike the acyclic cross-metathesis reactions, which generally favour the formation of *trans* products, the above ring-opening metathesis reactions yielded products in which the *cis* stereoisomer is predominant. Particularly noteworthy was the absence of significant amounts of products of type 31, formed from metathesis of one cyclic and two acyclic alkenes. In fact, considering the number of possible ring-opened products that could have been formed, these reactions showed remarkable selectivity (GC yields > 80%).

Ring-opening cross-metathesis of unsymmetrical cyclobutenes was also accomplished, although an extra complication arises due to the possible formation of two regioisomers (for example 39 and 40) of the desired cross-metathesis product (for example Eq. 32).

39: cis:trans = 2.3:1   40: cis:trans = 1.7:1

Unfortunately, the regioselectivity in this reaction was very poor. A better selectivity was observed, however, for a single example in which the cyclic alkene contained a substituent at the ring junction (4:1 mixture of regioisomers).

When analogous reactions were performed using symmetrical internal acyclic alkenes only polymerisation of the cyclobutene substrate was observed.

3.3.2
The Catalytic Cycle

Snapper proposed that the selectivity for the formation of cross-metathesis products 41 observed in these reactions was due to the differing reactivities of the various ruthenium alkylidene species formed in the catalytic cycle (Scheme 6).

The sterically bulky ruthenium alkylidene 42, formed via ring-opening of the cyclobutene, should react more rapidly with the terminal alkene than with a second molecule of the cyclobutene. This preference for reacting with the acyclic alkene is probably due to a combination of the greater steric hindrance of the cyclic alkene and the ability of the reaction with the terminal alkene to proceed

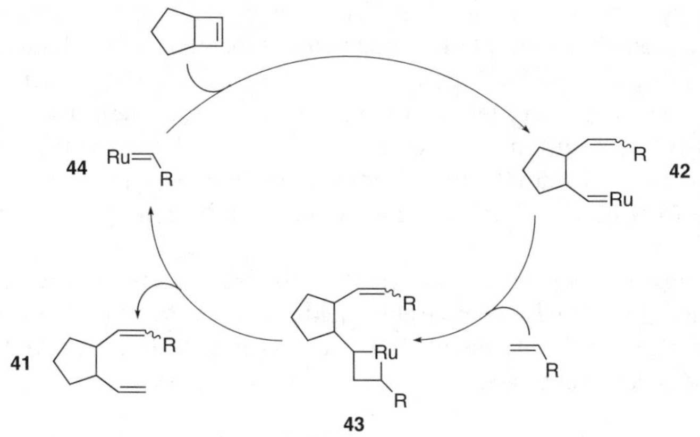

Scheme 6. Catalytic cycle leading to the formation of the desired cross-metathesis product **41**

through an intermediate α,α'-disubstituted metallacyclobutane 43. A preference in this step for proceeding via an α,α'-disubstituted, rather than an α,β-disubstituted metallacyclobutane, would explain why only ring-opened products containing a single R group are formed. The less hindered alkylidene 44 now formed should be less sensitive to steric factors and therefore would react preferentially with the highly reactive strained cyclobutene to reform 42. As Scheme 6 shows, this would result in selective formation of the desired cross-metathesis product 41. Further evidence for this catalytic cycle was provided in a more recent publication by Snapper and co-workers [48].

3.4
Ring-Opening Cross-Metathesis of Norbornenes

3.4.1
Reactions with Internal Acyclic Alkenes

Successful ring-opening cross-metathesis with symmetrical internal acyclic alkenes was, however, achieved by Blechert and Schneider [49]. Reaction of a variety of functionalised norbornene derivatives with *trans*-hex-3-ene in the presence of the ruthenium vinylalkylidene catalyst 4 yielded the ring-opened products as predominantly *trans-trans* isomers (for example Eq. 33).

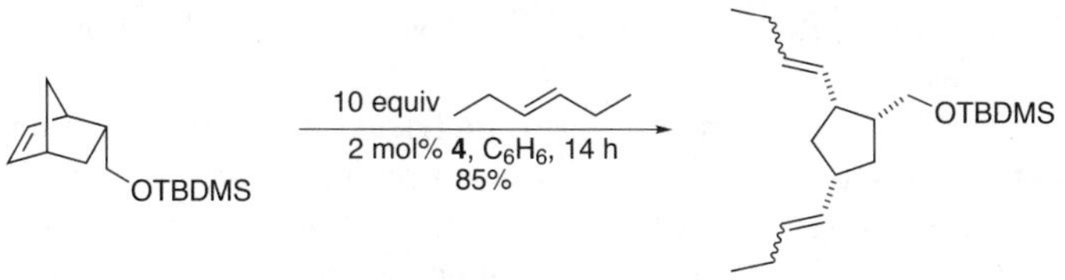

$$(33)$$

45: trans-trans:trans-cis = 2:1

Use of a symmetrical acyclic alkene limits the possible metathesis products to the desired diene (for example **45**) and products formed from polymerisation of the cyclic substrate. Competing ROMP was suppressed in these reactions by using dilute conditions and a tenfold excess of hex-3-ene. By adding the cyclic substrate slowly to a solution of the catalyst and *cis*-hex-3-ene (which was significantly more reactive than the *trans* isomer), less than two equivalents of the acyclic alkene were used without causing a significant drop in the cross-metathesis yield.

Replacing hex-3-ene with *trans*-1,4-dimethoxybut-2-ene resulted in slightly slower reactions, but gave comparable yields of cross-metathesis products. The desired reactions did not take place, however, when *cis*-but-2-ene-1,4-diol was used as the acyclic substrate.

3.4.2
Reactions with Terminal Acyclic Alkenes

A subsequent publication by Blechert and co-workers demonstrated that the molybdenum alkylidene **3** and the ruthenium benzylidene **17** were also active catalysts for ring-opening cross-metathesis reactions [50]. Norbornene and 7-oxanorbornene derivatives underwent selective ring-opening cross-metathesis with a variety of terminal acyclic alkenes including acrylonitrile, an allylsilane, an allyl stannane and allyl cyanide (for example Eq. 34).

cis:trans = 2:1

(34)

Unlike the previous reports, very good selectivities were obtained here using just 1–1.5 equivalents of the acyclic alkene without the need for slow addition of one of the substrates. With the exception of the reactions involving acrylonitrile or allyltributyl stannane, which used the molybdenum catalyst, all the reactions reported were catalysed by the Grubbs ruthenium benzylidene. Ring-opening cross-metathesis with very sterically hindered acyclic substrates, such as vinylsilanes and 1,1-disubstituted alkenes, proved to be unsuccessful. This was presumably due to competing ROMP of the cyclic substrate being the more facile process in these cases.

As mentioned earlier, when using an unsymmetrical cyclic substrate there is the possibility of forming regioisomers. Although, previously, only moderate regioselectivity had been observed with these reactions, the ring-opening cross-metathesis of lactam **46** with allyltrimethylsilane occurred with complete regioselectivity to yield diene **47** (Eq. 35).

$$(35)$$

46   **47**: trans:cis = 2:1

Ring-opening was not restricted just to norbornenes: bicyclo[3.3.0]octenes, cyclooctene and a functionalised cyclobutene were all reactive substrates for these reactions.

3.5
Regioselectivity Through Substrate Modification

3.5.1
Tricyclic Cyclobutenes

At the beginning of 1997, about the same time as Blechert reported the above metathesis reactions of norbornenes with terminal alkenes, Snapper and co-workers published the application of ring-opening cross-metathesis to the synthesis of bicyclo[6.3.0] ring systems [51]. Ruthenium catalysed ring-opening cross-metathesis of a functionalised cyclobutene **48** with ethene yielded the bicyclic diene **49**. A subsequent Cope rearrangement afforded the desired bicyclo[6.3.0] ring system **50** (Eq. 36).

$$(36)$$

48   **49**   **50**

Reaction of this same cyclobutene substrate **48** with a terminal alkene (TB-DMS protected pent-4-en-1-ol) gave a good yield (84%) of the cross-metathesis products, but with very little regioselectivity (3:2 mixture of regioisomers).

Good levels of regioselectivity were observed, however, when analogous cyclic substrates containing a hydroxy (**51**) or methyl substituent (**52**) projecting from the *exo*-face of the cyclobutene were used. Formation of exclusively *trans* double bonds in the major regioisomers was also observed with these substrates (Eq. 37) (Table 4).

$$(37)$$

51, 52   major regioisomer

Table 4. Cross metathesis reactions of **51** and **52** with silyl ether **17**

	X	R	R'	Yield (%)	Regioisomer ratio
51	CH_2	CO_2Me	OH	60	10:1
52	O	Pr	Me	72	8:1

The ring-opening cross-metathesis reaction of the diastereomer of **51**, in which the hydroxy group is on the opposite face, occurred with very little regioselectivity (1.9:1) and yielded the products as *cis/trans* mixtures. The different reactivities of the two diastereomers was even more pronounced in the methyl substituted cyclobutenes: no ring-opening metathesis occurred at all when the diastereomer of **52** (with the methyl group on the opposite face) was used. These results indicate that the regioselectivity observed here was due to the steric hindrance of the substituent rather than electronic effects or coordination to the ruthenium alkylidene.

3.5.2
Polymer-Bound Norbornenes

More recently, regioselective ring-opening metathesis has also been observed in reactions involving polymer-bound norbornene substrates [52]. Initial solution-phase ring-opening cross-metathesis reactions performed with *para*-methoxystyrene and functionalised unsymmetrical norbornenes gave high yields of the desired diene products, but little regioselectivity (Eq. 38). Syringe pump addition of the norbornene to the reaction mixture was required to minimise ROMP.

$$\text{5–10 mol\% 17, DCM} \qquad \text{61–90\%} \tag{38}$$

regioisomer ratio = 1.3:1–1.6:1

Ar = C_6H_4OMe

X = N-Boc piperazine   NHBu
     $NHCH_2C_6H_4OMe$ $NHCH_2CH_2CH_2NHBoc$

In order to avoid ROMP, ring-opening cross-metathesis reactions between a norbornene substrate (for example **53**), attached to Wang resin (0.85–1.01 mmolg^{-1}) via a variety of diamine linkers, and substituted styrenes were investigated. Use of a piperazine linker or electron deficient styrenes gave modest regioselectivities at best. Ring-opening metathesis reactions between electron rich styryl ethers and resin-bound norbornenes attached through primary diamine linkers, however, were completely regioselective (Eq. 39).

$$(39)$$

Ar = C$_6$H$_3$OR^{1}R^2

R^1 = Me, Ph, H; R^2 = H, OMe

X = CH$_2$CH$_2$CH$_2$ [CH$_2$]$_2$O[CH$_2$]$_2$O[CH$_2$]$_2$
CH$_2$-m-C$_6$H$_4$CH$_2$

No explanation for this selectivity was given, although it was noted that an analogous reaction using a TentaGel resin, containing a long poly(ethylene glycol) tether, was not completely regioselective.

4
Conclusion

The results discussed herein show that the cross-metathesis reaction is beginning to display considerable potential as a synthetic tool for intermolecular carbon–carbon bond formation. Using the Schrock molybdenum or Grubbs ruthenium carbene catalysts, an impressive array of functionalised alkenes have been successfully used as substrates in the cross-metathesis reaction. The tolerance of sulphide and phosphine groups exhibited by the tungsten alkylidene catalyst, developed by Basset, expands the range of possible substrates still further. Even substrates containing multiple functional groups such as β-lactams, α-amino acids and glycosides readily undergo cross-metathesis. Substrates with functional groups attached to the double bond (with the exception of acrylonitrile and vinylsilanes) do, however, pose a problem for the alkylidene catalysts currently available. The reaction is also currently limited to the formation of disubstituted double bonds.

Before the cross-metathesis reaction can be widely utilised in organic synthesis, however, the cross-/self-metathesis and *cis/trans* selectivity of the reaction need to be more reliable. Although chemo- and stereoselectivities of reactions with certain alkenes, such as styrene and acrylonitrile, are regularly very high and can be accurately predicted, substrates of this type are currently limited to a select few. The very good cross-/self-metathesis selectivity generally observed with sterically bulky alkenes, for example allyltrimethysilane, is a promising step, however, towards a more comprehensive method of directing chemoselectivity.

For the cross-metathesis of functionalised alkenes the ill-defined 'classical' catalyst systems currently offer very few advantages (cost and heterogeneous catalysis) over the more functional group tolerant Schrock and Grubbs alkylidene catalysts, particularly for laboratory scale organic synthesis. Although there are still some functionalised alkenes which display better cross-metathesis reactivity with the ill-defined catalysts, such as α,β-unsaturated esters, the number of examples is very small.

The cross-metathesis reaction has evolved extensively during the past few years, but there is still a considerable amount of work to be done before the full potential of this reaction is realised. The development of new metathesis catalysts, greater understanding and control of selectivity, and more extensive applications in synthesis that will surely follow in the near future, make this a particularly exciting time in the evolution of the alkene cross-metathesis reaction.

5
References

1. For a more in depth coverage of the use of ill-defined catalysts in cross-metathesis, see Ivin KJ, Mol JC (1997) Olefin metathesis and metathesis polymerisation. Academic Press, San Diego, Chap 9
2. van Dam PB, Mittelmeijer MC, Boelhouwer C (1972) J Chem Soc, Chem Commun 1221
3. Bosma RHA, van den Aardweg GCN, Mol JC (1981) J Chem Soc, Chem Commun 1132
4. Bosma RHA, van den Aardweg GCN, Mol JC (1983) J Organomet Chem 255:159
5. Marciniec B, Foltynowicz Z, Lewandowski M (1994) J Mol Catal 90:125
6. Finkel'shtein ES, Portnykh EB, Antipova IV, Vdovin VM (1989) Izv Akad Nauk SSSR, Ser Khim 38:1358
7. (a) Berglund M, Andersson C, Larsson R (1985) J Organomet Chem 292:C15. (b) Berglund M, Andersson C (1986) J Mol Catal 36:375
8. (a) Foltynowicz Z, Pietraszuk C, Marciniec B (1993) Appl Organomet Chem 7:539. (b) Foltynowicz Z, Marciniec B (1997) Appl Organomet Chem 11:667
9. Nakamura R, Echigoya E (1977) Chem Lett 1227
10. Warwel S, Döring N, Deckers A (1988) Fat Sci Techn 90:125
11. (a) Bosma RHA, Kouwenhoven AP, Mol JC (1981) J Chem Soc, Chem Commun 1081. (b) Bosma RHA, van den Aardweg GCN, Mol JC (1985) J Organomet Chem 280:115. (c) Bespalova NB, Bovina MA (1992) J Mol Catal 76:181
12. Bespalova NB, Bovina MA, Sergeeva MB, Oppengeim VD, Zaikin VG (1994) J Mol Catal 90:21
13. Quignard F, Leconte M, Basset J-M (1985) J Chem Soc, Chem Commun 1816
14. Schaverien CJ, Dewan JC, Schrock RR (1986) J Am Chem Soc 108:2771
15. Thu CT, Bastelberger T, Höcker H (1985) J Mol Catal 28:279
16. Schrock RR, Murdzek JS, Bazan GC, Robbins J, DiMare M, O'Regan M (1990) J Am Chem Soc 112:3875
17. Nguyen ST, Grubbs RH, Ziller JW (1993) J Am Chem Soc 115:9858
18. Couturier J-L, Paillet C, Leconte M, Basset J-M, Weiss K (1992) Angew Chem, Int Ed Engl 31:628
19. Lefebvre F, Leconte M, Pagano S, Mutch A, Basset J-M (1995) Polyhedron 14:3209
20. Couturier J-L, Tanaka K, Leconte M, Basset J-M, Ollivier J (1993) Angew Chem, Int Ed Engl 32:112
21. Descotes G, Ramza J, Basset J-M, Pagano S (1994) Tetrahedron Lett 35:7379
22. Leconte M, Jourdan I, Pagano S, Lefebvre F, Basset J-M (1995) J Chem Soc, Chem Commun 857
23. Fu GC, Grubbs RH (1992) J Am Chem Soc 114:5426
24. Fu GC, Grubbs RH (1992) J Am Chem Soc 114:7324
25. Crowe WE, Zhang ZJ (1993) J Am Chem Soc 115:10998

26. Fox HH, Schrock RR, O'Dell R (1994) Organometallics 13:635
27. Crowe WE, Goldberg DR (1995) J Am Chem Soc 117:5162
28. Brümmer O, Rückert A, Blechert S (1997) Chem Eur J 3:441
29. Crowe WE, Goldberg DR, Zhang ZJ (1996) Tetrahedron Lett 37:2117
30. Barrett AGM, Beall JC, Gibson VC, Giles MR, Walker GLP (1996) J Chem Soc, Chem Commun 2229
31. Barrett AGM, Baugh SPD, Gibson VC, Giles MR, Marshall EL, Procopiou PA (1996) J Chem Soc, Chem Commun 2231
32. Barrett AGM, Baugh SPD, Gibson VC, Giles MR, Marshall EL, Procopiou PA (1997) J Chem Soc, Chem Commun 155
33. It has been noted in a recent review that 4 successfully catalyses the cross-metathesis reactions of methyl oleate 1 or oleic acid with ethene; see [1]
34. Schwab P, France MB, Ziller JW, Grubbs RH (1995) Angew Chem, Int Ed Engl 34:2039
35. Schwab P, Grubbs RH, Ziller JW (1996) J Am Chem Soc 118:100
36. Catalyst 17 is commercially available from Strem Chemicals.
37. Schuster M, Pernerstorfer J, Blechert S (1996) Angew Chem, Int Ed Engl 35:1979
38. Schuster M, Lucas N, Blechert S (1997) J Chem Soc, Chem Commun 823
39. Feng J, Schuster M, Blechert S (1997) Synlett 129
40. (a) Huwe CM, Velder J, Blechert S (1996) Angew Chem, Int Ed Engl 35: 2376. (b) Fu GC, Nguyen ST, Grubbs RH (1993) J Am Chem Soc 115:9856
41. Fu GC, Grubbs RH (1993) J Am Chem Soc 115:3800
42. Gibson (née Thomas) SE, Gibson VC, Keen SP (1997) J Chem Soc, Chem Commun 1107
43. Biagini SCG, Gibson (née Thomas) SE, Keen SP (1998) J Chem Soc, Perkin Trans 1 0000
44. Pernerstorfer J, Schuster M, Blechert S (1997) J Chem Soc, Chem Commun 1949
45. Boger DL, Chai W, Ozer RS, Andersson C-M (1997) Bioorg Med Chem Lett 7:463
46. Feher FJ, Soulivong D, Eklund AG, Wyndham KD (1997) J Chem Soc, Chem Commun 1185
47. Randall ML, Tallarico JA, Snapper ML (1995) J Am Chem Soc 117:9610
48. Tallarico JA, Bonitatebus Jr PJ, Snapper ML (1997) J Am Chem Soc 119:7157
49. Schneider MF, Blechert S (1996) Angew Chem, Int Ed Engl 35:411
50. Schneider MF, Lucas N, Velder J, Blechert S (1997) Angew Chem, Int Ed Engl 36:257
51. Snapper ML, Tallarico JA, Randall ML (1997) J Am Chem Soc 119:1478
52. Cuny GD, Cao J, Hauske JR (1997) Tetrahedron Lett 38:5237

Recent Advances in ADMET Chemistry

D. Tindall, J.H. Pawlow, and K.B. Wagener

Acyclic diene metathesis, the condensation of terminal dienes to yield high polymer, has been found to be an extremely versatile reaction. Using the appropriate choice of catalyst, polymers containing a wide variety of functional groups have been synthesized. A unique aspect of ADMET is its ability to produce new polymer backbones by strategic monomer design. ADMET has been utilized to make segmented block copolymers, metal containing polymers, and regularly branched polyolefins, that are difficult to synthesize by other means. New aspects of this chemistry are outlined, along with a discussion of the ADMET reaction mechanism, catalysts, and kinetics.

Keywords: Acyclic diene metathesis polymerization, ADMET, Condensation polymerization Functionalized polymers, Negative neighboring group effect, Branched polyethylene

1
Introduction

It is well documented that the productive metathesis exchange reaction of two acyclic olefins produces two new olefins; yet the reaction occurs efficiently with no obvious driving force except for the entropic increase associated with an equilibrium process [1]. However, if the newly created olefins are removed selectively from the reaction, then the productive metathesis equilibrium can be shifted decisively towards product formation. We have exploited this equilibrium reaction, developing the chemistry which is now known as acyclic diene metathesis (ADMET) polymerization [2]. The ADMET reaction has been extensively studied for several years to explore its viability in producing both pure hydrocarbon polymers as well as polymers possessing functional groups. Factors influencing the polymerization of both classes of monomers have been examined in detail, and it is now known that a wide variety of macromolecules can be produced by taking into account steric and electronic factors.

1.1
Mechanistic Aspects of the ADMET Reaction

The ADMET reaction is shown below in Eq. 1, which generically displays an α,ω-acyclic diene being condensed into its requisite generic polymer plus a small molecule.

$$\overset{}{\diagup}R\overset{}{\diagdown} \xrightleftharpoons{\text{Catalyst}} \left(\overset{}{\diagup}R\overset{}{\diagdown}\right)_n + CH_2=CH_2 \qquad (1)$$

Terminal dienes are the preferred monomers in the ADMET reaction, due to both entropic and steric considerations, which will be discussed in greater detail

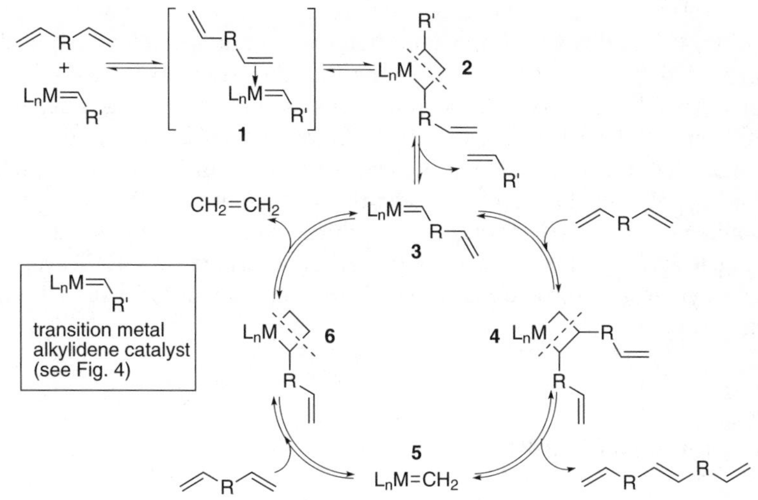

Scheme 1. The mechanism for ADMET polymerization

later in this chapter. The small molecule formed during metathesis of terminal olefin units is ethylene, which is initially removed from the reaction by a simple change in state of matter to the gaseous phase and then can be completely removed from the system in vacuo. Internal olefin sites can also undergo comparable metathesis chemistry, concurrently forming the appropriate alkene but at a cost of reduced activity and rate. Another significant factor to contemplate is that ADMET chemistry is catalyst dependent, for it is known that other mechanisms can compete with the metathesis reaction to an extent that obviates the generation of soluble high molecular weight polymers. Catalyst selection has been limited to transition metal complexes that are essentially free of Lewis acids, which have been shown to favor vinyl addition over metathesis chemistry [3]. However, if the appropriate acid-free catalysts are employed then the ADMET mechanism predominates, as shown in Scheme 1 [4].

The ADMET mechanism is characterized by the presence of a metallacyclobutane ring, a reaction intermediate which is found both in ring opening metathesis polymerization (ROMP) chemistry as well as ring closing metathesis (RCM) chemistry [1, 5]. It has been conclusively demonstrated that this metallacyclobutane ring is generated during the ADMET reaction, and in fact is found in a number of places in the reaction cycle which is illustrated above [6]. The initial step is the formation of a π complex, **1**, between one of the olefin groups of the diene and the electron deficient metal center, followed by collapse to a metallacyclobutane ring, **2**. In productive metathesis this ring cleaves in such a manner as to place the metal center on the end of monomer, and this species, **3**, initiates the polymerization cycle. The next event is analogous to the one which has just occurred; one of the olefins in a diene or in a polymer chain end forms a π complex to the electron deficient metal center, which leads to another metallacyclobutane ring, **4**. It is this very important metallacyclobutane ring which leads to polymer creation, for it results in the formation of a larger molecule via the connection of two diene units. The likelihood of this event occurring is dependent on both steric and electronic features which are described later in this chapter. In the next step of the mechanism, **4** collapses to form an internal olefin within the growing polymer chain and generates the true catalyst in ADMET chemistry, a methylidene, **5**. The cycle continues by a third formation of a metallacyclobutane ring, **6**, again via the same π complexation described earlier. Cleavage of **6** evolves ethylene, and the cycle is repeated, growing polymer in a stepwise fashion with every cycle.

1.2
Comparison of ADMET With RCM

It is evident from this mechanistic illustration that the reactions which do occur must be highly efficient and precise in their nature; otherwise this equilibrium step-growth process could never be successful in producing high molecular weight molecules. In this respect, it is clear that ADMET chemistry functions in a manner very similar to ring closing metathesis (RCM) chemistry [1]. Neither of these two metathesis processes is favorable unless the metathesis chemistry itself occurs to completion, for virtually quantitative conversions are needed in both cases to be useful in synthetic chemistry. Essentially, the ADMET reaction (Eq. 2) and the RCM reaction (Eq. 3) are inter- and intramolecular displays of the same mechanistic event. In the RCM reaction, intramolecular metathesis closes a ring to form a small cyclic molecule with concurrent loss of a small molecule. Conversely, in the case of the ADMET reaction, macromolecules are produced via successive intermolecular condensation of two olefins. Thus, the information that is presented in subsequent sections within this chapter for the formation of ADMET polymers has a direct bearing on many of the reaction features involved in RCM chemistry.

$$\text{R} \quad + \quad L_nM = \Big\backslash_{R'} \quad \xrightarrow[- \ CH_2=CH_2]{\text{ADMET}} \quad \Big(\text{R} \Big)_n \tag{2}$$

$$\text{R} \quad + \quad L_nM = \Big\backslash_{R'} \quad \xrightarrow[- \ CH_2=CH_2]{\text{RCM}} \quad \Big(\text{R} \Big) \tag{3}$$

The balance of this chapter deals with the specific chemistry associated with producing hydrocarbon and functionalized polymers in addition to providing the most recent studies available on appropriate catalyst systems for ADMET condensation chemistry. Current work on the use of the ADMET reaction for modeling commercial high volume polymers such as polyethylene is also presented.

2
Preparation of Hydrocarbon and Functionalized ADMET Polymers

2.1
Saturated Hydrocarbon Dienes

The outset of ADMET research began with an examination of pure hydrocarbon dienes that could be converted to their corresponding high molecular weight step polymerization polymers. The first monomer chosen in this study was 1,9-decadiene, and by using an acid-free catalyst, we were able to convert it with relative ease to a high molecular weight polymer with a number average molecular weight value of 50,000 [2, 7]. The reaction was carried out under bulk polymerization conditions to maximize monomer concentration in this equilibrium reaction with monomer to catalyst ratios of roughly 1000:1. Conversions occurred over a period of hours leading to polymers exhibiting polydispersity indices approaching 2.0 [6]. These results are typical for any equilibrium step polymerization, such as those found in the formation of polyamide or polyester, and subsequent work on the ADMET reaction has shown it is comparable to any polycondensation reaction that is known today.

2.2
Sterically Encumbered Monomers

Moving methyl groups in succession towards internal carbons in the diene illustrates how steric effects play a role in the ADMET reaction. When a methyl group is present on the internal sp^2 hybridized carbon, as in monomer 7, the ADMET reaction ceases (Eq. 4) [8]. This can be explained mechanistically by examining the reaction cycle shown in Scheme 1 where formation of the second metallacyclobutane ring, 4, is prevented due to steric influences at this position. Equation 5 shows a further illustration of this point, where removing one of the methyl groups results in simple dimerization of 8 since one olefin in the monomer remains accessible while the other olefin site is sterically rendered inactive. Subsequent studies of this nature have shown that this steric effect can even be important at the position α to the double bond [9]. Sterically encumbering this position also hinders formation of the necessary metallacyclobutane ring, inhibiting polymer formation.

$$(4)$$

$$(5)$$

2.3
Tailored Polymers from Diene Monomers

A number of hydrocarbon polymer structures have been made using the AD-MET reaction when taking into consideration the steric effects that have been previously described. For example, it is possible to produce both saturated [10] and conjugated [11] hydrocarbon polymers by ADMET chemistry, resulting in the formation of very pure polymers. Typical examples include poly(phenylene vinylene) [12], 1,4-polybutadiene [13], and polyoctenamer [14], polymers which are of general interest in the materials community today. Furthermore, as shown in Eq. 6, it is possible to produce polymers with a microstructure identical to perfectly alternating copolymers from a single monomer by choosing a triene monomer, such as **9** (Eq. 6) [15]. In this monomer, the internal, sterically encumbered olefin does not participate in the ADMET reaction, thereby resulting in a polymer repeat unit which perfectly alternates isoprene and 1,4-polybutadiene.

$$\text{9} \qquad \xrightarrow{\text{Catalyst}} \qquad \left[\right]_n \qquad (6)$$

2.4
ADMET Depolymerization

In addition, it is interesting to note that the polymerization of hydrocarbon monomers is completely reversible [16]. We have demonstrated that unsaturated polymers such as 1,4-polybutadiene can be converted to diene monomers via depolymerization with ethylene [17]. These conversions, illustrated in Eq. 7, show that it is possible not only to achieve high mass conversions of the polymer to 1,5-hexadiene, **11**, but it is also possible to create telechelic materials, **10**, in this manner [18]. Catalyst selection plays an important role here, and the most recent work shows that ruthenium catalysts are best in bringing about clean conversions of high molecular weight unsaturated polymers to their telechelic oligomers. Furthermore, it has been shown that one can incorporate functional groups such as alcohols, esters, carboxylic acids, and imides into α,ω-telechelic dienes by ADMET depolymerization [19]. These telechelic oligomers can then be used in further reactions to create materials such as hydrophobic polyurethanes and other specialty polymers.

$$\left[\right]_n \quad \xrightarrow{CH_2=CH_2} \quad \left[\right]_m \quad \xrightarrow{CH_2=CH_2} \quad \qquad (7)$$
polybutadiene $\qquad\qquad\qquad$ **10** $\qquad\qquad\qquad$ **11**

2.5
Functionalized Diene Monomers

Applying this methodology to monomers possessing functional groups, it is now evident that a wide variety of functionalized, unsaturated and saturated polyolefins can be prepared using ADMET chemistry. Catalyst choice as well as position of the functional group within the monomer makes a significant impact, both with respect to rate of polymerization and molecular weights achieved. The optimum conditions for ADMET chemistry occur by positioning the functional group at least two methylene units distant from the metathesizing olefin within the monomer unit. If the functional group is any closer, an intramolecular complex can form between the lone pair of electrons within the functional group and the electron-deficient metal center. This phenomenon, illustrated in Fig. 1, has been termed the "negative neighboring group effect" [20].

Referring to the ADMET mechanism discussed previously in this chapter, it is evident that both intramolecular complexation as well as intermolecular π-bond formation can occur with respect to the metal carbene present on the monomer unit. If intramolecular complexation is favored, then a chelated complex, **12**, can be formed that serves as a thermodynamic well in this reaction process. If this complex is sufficiently stable, then no further reaction occurs, and ADMET polymer condensation chemistry is obviated. If in fact the chelate complex is present in equilibrium with π complexation leading to a polycondensation route, then the net result is a reduction in the rate of polymerization as will be discussed later in this chapter. Finally, if **12** is not kinetically favored because of the distant nature of the metathesizing olefin bond, then its effect is minimal, and condensation polymerization proceeds efficiently. Keeping this in perspective, it becomes evident that a wide variety of functionalized polyolefins can be synthesized by using controlled monomer design, some of which are illustrated in Fig. 2.

All of these monomers can produce polycondensation materials of typical number average molecular weights between 10,000 and 30,000. Molecular weight distributions are 2.0, typical in any ADMET reaction using hydrocarbon dienes, and the polymers offer a great diversity in terms of response, depending on the nature of the functional group which is incorporated. We have been able to prepare polyethers [20b, 21], polycarbonates [22], polyesters [23], polyketones [24], polysiloxanes [25], polycarbosiloxanes [26], polycarbodichlorosilanes [27], etc., in a manner that shows that the only limiting factor in this reac-

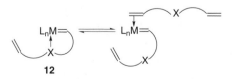

12

Fig. 1. The negative neighboring group effect

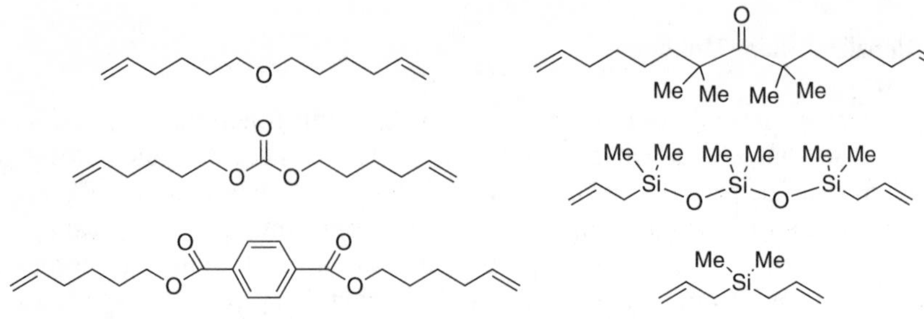

Fig. 2. Some representative functionalized diene monomers

tion is the ease with which the monomer dienes themselves can be prepared. At this point, we expect that virtually any functional group could be included in an ADMET polymer if in fact the proper spacing is placed between the metathesizing double bond and the functional group itself.

2.6
Segmented Copolymers by ADMET

It has been possible to design a series of segmented copolymers in which functionality can be present either in the hard segment, the soft segment, or in both, as is desired. Soft segment α,ω-dienyl telechelic poly(tetrahydrofuran) has been synthesized and shown to be polymerizable under ADMET conditions [28]. When copolymerized with 1,9-decadiene or ester functionalized diene monomers, segmented copolymers are produced, as shown in Eq. 8. The poly(tetrahydrofuran) segment serves as the "soft" phase and the polyoctenamer or polyester unit serves as the "hard" phase in the resulting material. The development of α,ω-dienyl telechelic oligomers composed of other repeat units is underway (Fig. 3), and it is anticipated that as long as these species meet the steric and electronic restrictions of the ADMET reaction, there should be little restriction to their application as effective ADMET monomers. Drawing from the broad range of functionalized diene monomers known to be active in ADMET polymerization, an array of segmented copolymers with tailored properties is under investigation, where the unique aspects of ADMET chemistry enable the synthesis of these polymers which are not easily accessible by other synthetic means.

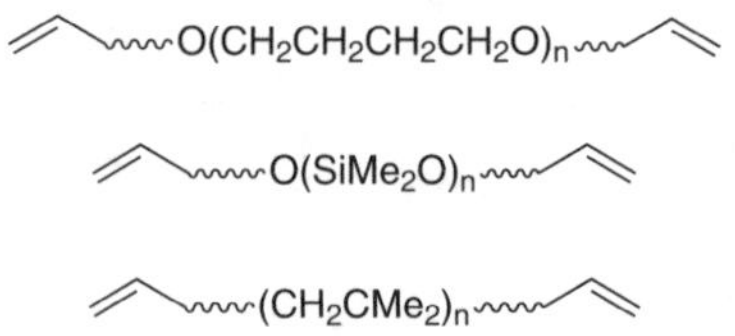

Fig. 3. α,ω-Dienyl telechelic oligomers used for producing segmented copolymers by ADMET

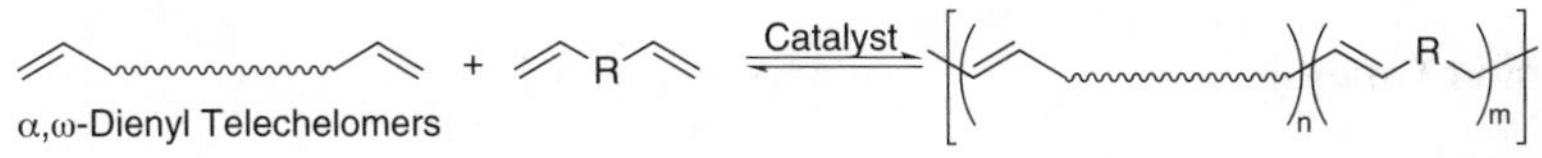
α,ω-Dienyl Telechelomers
Catalyst
R
n
m

(8)

3
Recent Developments in ADMET Catalysis

In condensing dienes into new polymers, the most significant observation on the ADMET reaction concerns the acidity effect in the catalyst used in bringing about this conversion. Highly acidic catalysts induce the competition of vinyl addition chemistry, preventing high conversions of dienes to their respective condensation polymers [1–4]. As a consequence, research over the past 10 years has focused on catalyst systems free of acids, and three such systems have been found to be useful in producing polymers of high molecular weight [29–31]. These three catalytic systems are illustrated in Fig. 4.

3.1
Tungsten and Molybdenum Catalysis

ADMET chemistry commenced using the tungsten version of Schrock's molybdenum catalyst, **13a**. This and other highly hindered complexes based on both tungsten and molybdenum are the fastest ADMET catalysts known to date, producing polymers at a reaction rate comparable to that of acid chloride-alcohol polycondensation chemistry [6, 7, 32]. Rapid conversions can be performed with virtually any type of ADMET-active monomer, even those possessing heteroatoms. Additionally, these catalyst systems promote very clean conversions of substrate. The nature of all the chemical species involved in the reaction cycle has been identified and characterized by a variety of spectroscopic means. These polymerization systems behave analogously to any other polycondensation reaction in the presence of an active catalyst system. Molecular weights of the products can be increased by further condensation and, as noted earlier, the reaction can be reversed by addition of the small molecule which has been eliminated as a result of the polycondensation event.

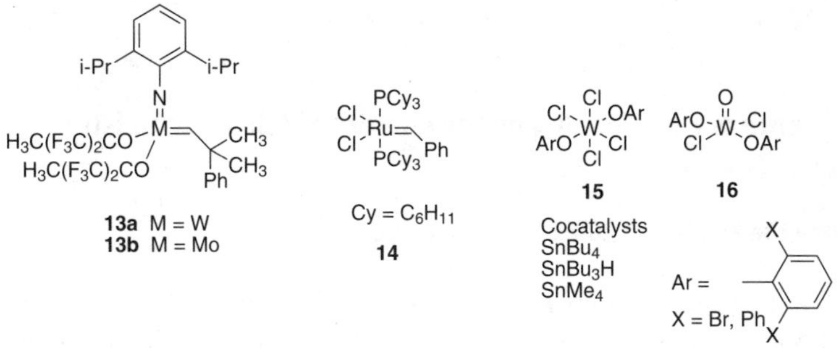
i-Pr
i-Pr
N
H3C(F3C)2CO
H3C(F3C)2CO
M
CH3
CH3
Ph
13a M = W
13b M = Mo
Cl
PCy3
Ru
Ph
Cy = C6H11
14
OAr
ArO
W
15
Cocatalysts
SnBu4
SnBu3H
SnMe4
O
16
Ar =
X = Br, Ph
X

Fig. 4. ADMET catalysts

3.2
Ruthenium Catalysis

The ruthenium catalyst system, **14**, shown in Fig. 3, also carries out ADMET condensation chemistry, albeit with higher concentrations being required to achieve reasonable reaction rates [32]. The possibility of intramolecular complexation with this catalyst influences the polymerization reaction, but nonetheless, ruthenium catalysis has proved to be a valuable contributor to overall condensation metathesis chemistry. Equally significant, these catalysts are tolerant to the presence of alcohol functionality [33] and are relatively easy to synthesize. For these reasons, ruthenium catalysis continues to be important in both ADMET and ring closing metathesis chemistry.

3.3
Classical Catalyst Systems

Most recently it has been demonstrated that classical metathesis catalyst systems such as those shown above are capable of inducing ADMET condensation chemistry [34]. These classical systems, **15** and **16**, are precursors to actual metal carbenes, and they must be activated with the presence of an alkylating agent such as tetrabutyltin or tributyltin hydride. The ADMET condensation chemistry proceeds at a reasonable rate and high molecular weight polymers can be obtained.

Exploiting this latter observation, we have created monomers which also serve as cocatalysts in classical metathesis reactions, thereby producing a "self-polymerizing" monomer system [35]. As shown in Eq. 9, bis(4-pentenyl)dibutylstannane condenses to its ADMET polymer in the presence of **15** or **16** without the need of added cocatalyst. This conversion occurs at a reasonably high rate and produces a clean metal-containing polymer possessing roughly 35 weight % tin in its backbone [36]. Monomers that can function as cocatalysts are of interest in terms of catalyst activation, and this approach is the focus of current research.

$$\text{(9)}$$

4
The Influence of Functionality on the Kinetics of ADMET Chemistry

4.1
Measuring the Kinetics of ADMET Polymerization

While the effects of various functionalities and their spatial location on ADMET chemistry are now evident, only recently has the dramatic influence of the functional group on the kinetics of the reaction depending upon the organometallic catalyst chosen been studied quantitatively [20a]. It is well known that the rate of ring opening metathesis polymerization can be mediated by the presence of Lewis base containing solvents. The complex formed between the active catalyst and the base dilutes the effective concentration of catalyst available to carry out the polymerization chemistry desired. In order to examine this phenomenon within the ADMET reaction scheme, a series of dienes were prepared possessing both oxygen and sulfur atoms in which the heteroatom has been placed at least three carbons distant from the metathesizing double bond.

Two well characterized and highly efficient catalyst complexes were chosen for this study, Schrock's molybdenum catalyst, **13b** ($Mo(NC_6H_3$-2,5-i-Pr)(OC $(CH_3)(CF_3)_2)_2(CHC(CH_3)_2C_6H_5))$, and Grubbs's ruthenium catalyst, **14** ($RuCl_2$ $(P(C_6H_{11})_3)_2CHC_6H_5)$. Kinetic data were generated by measuring the rate of ethylene evolution with time, and in all cases in which a reaction was observed the rate of polymerization was shown to be second order with respect to monomer [20a].

4.2
Influence of a Lewis Base Functional Group

Representative data illustrating the influence of Lewis base functional groups in the ADMET reaction are shown in Table 1. When molybdenum catalysts are used to polymerize ether or thioether dienes, little change in reaction rate is observed as compared with the standard, 1,9-decadiene, which possesses no heteroatoms in its structure. When a sulfur atom is three carbons atoms away from the alkene site, the reaction rate is reduced approximately one order of magnitude; otherwise, the kinetics are all essentially unaffected [20a].

However, using ruthenium catalysis it becomes immediately evident that the presence of a heteroatom significantly slows the reaction rate [32]. For example, an oxygen atom placed three carbons away from the reactive site reduces metathesis rates by approximately two orders of magnitude. Reactivity improvements are observed when the oxygen atom is placed four carbons away, suggesting that the negative neighboring group effect does play a role in the reaction. Furthermore, the presence of sulfur in the monomer terminates ruthenium catalyzed reactions altogether, regardless of whether it is three carbons or four car-

Table 1. Rate constants showing the kinetic effect of functionalized monomers with molybdenum and ruthenium catalysis

Diene Monomers	Catalysts	
	13b	14
	$2.4 \times 10^{-3}\,Lmol^{-1}s^{-1}$	$1.0 \times 10^{-4}\,Lmol^{-1}s^{-1}$
	$1.2 \times 10^{-3}\,Lmol^{-1}s^{-1}$	$6 \times 10^{-6}\,Lmol^{-1}s^{-1}$
	$2 \times 10^{-4}\,Lmol^{-1}s^{-1}$	0

bons away from the metathesizing olefin. Apparently the sulfur displaces the co-ordinated tricyclohexylphosphine ligands to form a stable intramolecular complex between the sulfur and the ruthenium center. Once this event has occurred, the stable complex formed is unreactive toward further metathesis activity, and it is unable to produce polymers of any significant molecular weight.

4.3
Mechanistic Differences With 1,5-Hexadiene

Using 1,5-hexadiene, it was shown that depending upon whether ruthenium or molybdenum catalysts are used, a change in mechanism appears to occur.

$$\text{(structure)} \xrightleftharpoons{\quad 13b \quad} \text{(structure)}_n + CH_2=CH_2 \qquad (10)$$

$$\text{(structure)} \xrightleftharpoons{\quad 14 \quad} \text{(structure)} + CH_2=CH_2 \qquad (11)$$

With molybdenum-based catalysis, 1,5-hexadiene produces principally line-ar polymers (Eq. 10) [20a]. A small quantity of cyclic species are formed in this conversion, but no more than is observed in a typical polycondensation reaction. However, when the ruthenium catalyst is used only minor quantities of ol-igomers are formed, with the primary product being the cyclic dimer, 1,5-cy-clooctadiene (Eq. 11) [20a, 32b]. This change in mechanism is attributed to a negative neighboring group effect, not with a heteroatom, but with the second olefin present in the diene. It is suspected that the terminal olefin forms a π chelate complex with the catalyst center, which then favors the intramolecular ring forming reaction over the intermolecular polymerization. Significant mass spectral evidence indicates that the formation of polymers from monomers con-taining either heteroatoms or closely spaced diene units is contingent upon the catalyst chosen.

5
Studying the Crystallization Behavior of Polyethylene

5.1
Modeling Commercial Polymers

With the details associated with ADMET chemistry reasonably well understood, we have embarked on a study of the synthesis of well-controlled polymer struc-tures via metathesis polycondensation chemistry [37]. A series of well-defined polyolefins have been designed to model the crystallization behavior of polyeth-ylene and its related copolymers, including new materials synthesized by metal-locene-based catalysts. This synthesis concept has been reduced to practice, and polymers that will aid in the understanding of branching within polyethylene it-self have been produced.

Branching in polymers plays a direct role in their crystallizability, as packing configurations are directly influenced by the nature and frequency of branches along a given polymer backbone. Much has been done in the past to understand the effect of randomly placed branches on the chain, yet no study has been performed on perfectly controlled branching in polyethylene.

5.2
Synthesis of Polyethylene by ADMET

We have synthesized such a material, which is called perfectly imperfect polyethylene, where each branch is a methyl group and its frequency along the backbone is controlled by the nature of the symmetrical diene used in the ADMET polycondensation reaction [37]. Equation 12 illustrates the chemistry used to produce polyethylene by a step polycondensation route rather than a chain propagation mechanism.

$$\text{(12)}$$

17	18	19
17a R = CH$_3$, n = 3	**18a** R = CH$_3$, n = 3	**19a** R = CH$_3$, n = 3

In this case, symmetrical substituted dienes are condensed to their respective ADMET polymers followed by hydrogenation to give a completely saturated polymer backbone. This technique gives control over the number of methylenes between branch points as well as the length and identity of the branch itself. However, synthesis of the required monomer **17** is a challenging six-step procedure.

ADMET condensation of **17** is completed using molybdenum catalysis to give the unsaturated polymer **18**, which is reduced to **19** using a variant of hydrazine reduction chemistry. Complete saturation of the polymer backbone has been demonstrated and is illustrated by the absence of olefin protons in the ^{13}C NMR of **19a** shown in Fig. 5.

5.3
Behavior of ADMET Polyethylene

Due to the symmetry of **19a**, which possesses a methyl group on each and every ninth carbon, the carbon NMR spectrum displays only six resonances. This conclusively demonstrates that these techniques are able to prepare polymers whose regularity can be controlled with ease. The polymers themselves are sufficiently high in molecular weight to mimic crystallization behavior of polyethylene in its substituted forms.

In the case of **19a**, it was found that the typical melting point for linear polyethylene, 134°C, with a heat diffusion of 204 Jg^{-1}, is depressed to –2°C with a heat diffusion of 32 Jg^{-1}. A typical DSC curve is shown in Fig. 6.

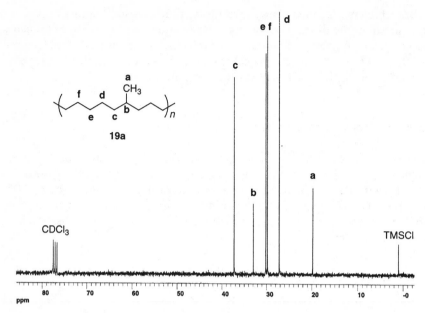

Fig. 5. ^{13}C NMR spectrum of methyl-branched polyethylene generated by ADMET

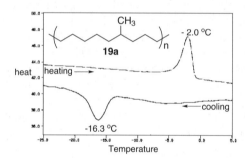

Fig. 6. DSC thermogram of methyl-branched polyethylene generated by ADMET

The nature of this type of crystal remains unknown and is the subject of further study.

This is just the first example of how the ADMET reaction can be used to model branching behavior and precisely control the structure in olefin-based polymer backbones. Other polymers under study include polyalcohols, polyvinyl acetates, and ethylene-styrene copolymers. The ultimate goal of this research is to be able to define, or even predict, crystallization limits and behavior for many polymers, some of which have not yet been prepared in a crystallized form.

6
Conclusions

The ADMET reaction is now clearly defined and has been shown to be a useful method for condensing a variety of hydrocarbon and functionalized dienes to their respective polymers. The depolymerization metathesis reaction has also been shown to be effective in producing telechelic oligomers. Catalyst and monomer development remain subjects of interest where the goal is to produce macromolecules by easily accessible and less expensive means. Further, acyclic diene metathesis polymerization is now being used to produce materials with well-defined backbone structures and architectures, such as perfectly alternating and segmented copolymers, and perfectly branched polyethylene. These materials contribute to the study of structure-property relationships such as the crystallization behavior of polymers in general.

7
References

1. For a comprehensive review of metathesis chemistry, see Ivin KJ, Mol JC (1997) Olefin metathesis and metathesis polymerization, Academic, San Diego, London and references therein
2. Lindmark-Hamberg M, Wagener KB (1987) Macromolecules 20:2949
3. Wagener KB, Nel JG, Konzelman J, Boncella JM (1990) Macromolecules 23:5155
4. Wagener KB, Boncella JM, Nel JG, Duttweiler RP, Hillmyer MA (1990) Makromol Chem 191:365
5. Ofstead EA, Wagener KB (1992) Polymer synthesis via metathesis chemistry. In: Mijs WJ (ed) New methods for polymer synthesis. Plenum, New York, Chap 8
6. Wagener KB, Boncella JM, Nel JG (1991) Macromolecules 24:2649
7. Wagener KB, Nel JG, Duttweiler RP, Hillmyer MA, Boncella JM, Konzelman J, Smith Jr DW, Puts RD, Willoughby L (1991) Rubber Chem Technol 64(1):83
8. (a) Konzelman J, Wagener KB (1995) Macromolecules 28:4686. (b) Wagener KB, Konzelman J (1991) Polym Prepr Am Chem Soc Div Polym Chem 32(1):375
9. Konzelman J, Wagener KB (1992) Polym Prepr Am Chem Soc Div Polym Chem 33(1):1072
10. O'Gara JE, Wagener KB, Hahn SF (1993) Makromol Chem 14:657
11. Tao D, Wagener KB (1994) Macromolecules 27:1281
12. Wolf A, Wagener KB (1991) Polym Prepr Am Chem Soc Div Polym Chem 32(1):535
13. Nel JG, Wagener KB, Boncella JM (1989) Polym Prepr Am Chem Soc Div Polym Chem 30(2):130
14. Nel JG, Wagener KB, Boncella JM, Duttweiler RP (1989) Polym Prepr Am Chem Soc Div Polym Chem 30(1):283
15. Konzelman J, Wagener KB (1996) Macromolecules 29:7657
16. Wagener KB, Nel JG, Smith Jr DW, Boncella JM (1990) Polym Prepr Am Chem Soc Div Polym Chem 31(2):711
17. (a) Wagener KB, Wolfe PS, Watson MD (1998) Metathesis depolymerization chemistry as a means of recycling polymers to telechelics and fine organic chemicals. In: Imamoglu Y (ed) Metathesis polymerization of olefins and polymerization of alkynes. Kluwer, Amsterdam, p 309. (b) Watson MD, Wagener KB (1996) Polym Prepr Am Chem Soc Div Polym Chem 37(1):609. (c) Wagener KB, Puts RD, Smith Jr DW (1991) Makromol Chem Rapid Commun 12(7):419 (d) Wagener KB, Puts RD (1991) Polym Prepr Am Chem Soc Div Polym Chem 32(1):379

18. (a) Marmo JC, Wagener KB (1995) Macromolecules 28:2602. (b) Marmo JC, Wagener KB (1993) Macromolecules 26:2137. (c) Wagener KB, Marmo JC (1995) Makromol Chem Rapid Commun 16:557
19. (a) Marmo JC, Wagener KB (1994) Polym Prepr Am Chem Soc Div Polym Chem 35(1):817. (b) Viswanathan T, Gomez F, Wagener KB (1994) J Polym Sci Part A Polym Chem 32:2469
20. (a) Wagener KB, Brzezinska K, Anderson JD, Younkin TR, DeBoer W (1997) Macromolecules 30:7363. (b) Brzezinska K, Wagener KB (1991) Macromolecules 24:5273
21. (a) O'Gara JE, Portmess JD, Wagener KB (1993) Macromolecules 26:2837. (b) Wagener KB, Brzezinska K, Bauch CG (1992) Makromol Chem 13:75
22. Wagener KB, Patton JT (1993) Macromolecules 26:249
23. (a) Wagener KB, Patton JT, Boncella JM (1992) Macromolecules 25:3862. (b) Bauch CG, Boncella JM, Wagener KB (1991) Makromol Chem Rapid Commun 12(7):413
24. Wagener KB, Patton JT, Forbes MD, Myers TL, Maynard HD (1993) Polym Int 32:411
25. (a) Smith Jr DW, Wagener KB (1993) Macromolecules 26:3533. (b) Smith Jr DW, Wagener KB (1991) Macromolecules 24:6073
26. Smith Jr DW, Wagener KB (1993) Macromolecules 26:1633
27. Cummings SK, Smith Jr DW, Wagener KB (1995) Makromol Chem Rapid Commun 16:347
28. Wagener KB, Brzezinska K, Anderson JD, Dilocker S (1997) J Polym Sci Part A Polym Chem 35:3441
29. Schrock RR, Murdzek JS, Bazan GC, Robbins J, Di Mare M, O'Regan M (1990) J Am Chem Soc 112:3875
30. Schwab PF, Marcia B, Ziller JW, Grubbs RH (1995) Angew Chem Int Ed Engl 34:2039
31. Nugent WA, Feldman J, Calabrese JC (1995) J Mol Catal 36:13
32. (a) Brzezinska K, Anderson JD, Wagener KB (1996) Polym Prepr Am Chem Soc Div Polym Chem 37(1):327. (b) Brzezinska K, Wolfe PS, Watson MD, Wagener KB (1996) Macromol Chem Phys 197:2065
33. Valenti DJ, Wagener KB (1996) Polym Prepr Am Chem Soc Div Polym Chem 37(1):325
34. Gomez FJ, Wagener KB (1997) PMSE Prepr Am Chem Soc Div Polym Mat Sci Eng 76:59
35. Wolfe PS, Gomez FJ, Wagener KB (1997) PMSE Prepr Am Chem Soc Div Polym Mat Sci Eng 76:250
36. Wolfe PS, Gomez FJ, Wagener KB (1997) Macromolecules 30:714
37. Wagener KB, Valenti D (1997) Macromolecules 30:6688

Bioactive Polymers

Laura L. Kiessling, Laura E. Strong

The use of synthetic polymers for biological applications is a rapidly growing field of research, and neobiopolymers have been found to elicit significant and new biological activities. The technology for the generation of these new materials has progressed considerably in the past few years. One method for neobiopolymer synthesis that provides a means to assemble materials containing the diversity of functional groups found in biological systems is the ring-opening metathesis polymerization (ROMP). This review will focus on the advances and advantages of ROMP for the generation of such neobiopolymers. To date, a prominent biological application of materials generated by ROMP has been in the exploration of the underlying features governing protein-carbohydrate interactions. Results in this area highlight the unique features of ROMP and point to many additional applications for this technology for the investigation and manipulation of biological systems.

Keywords: Polymers, Organometallic catalysts, ROMP, Multivalent

1
Introduction

Biopolymers play critical roles in the growth and development of all living things. The term biopolymers describes natural polymers such as proteins, DNA and RNA, and polysaccharides. Specific recognition of these biopolymers coordinates fundamental physiological interactions. The generation of synthetic biopolymers, which we refer to as neobiopolymers, that mimic natural biopolymers could have profound applications in disease prevention and treatment, as well as in the elucidation of biological mechanisms. These applications are predicated on the development of new methods for the controlled synthesis of these materials.

There are several important features of natural biopolymers that must be considered in the design of synthetic biopolymer mimics. First the backbone must display the necessary functional groups in orientations that are similar to those found in natural displays. In addition, the final products of these syntheses should be compatible with all the relevant environments in which the natural biomolecules occur. Finally, many biopolymers are homogeneous, an important feature in eliciting and elucidating specific responses. The generation of linear displays, or linear polymers, which share the structural features of natural biopolymers, is a rapidly expanding area. Many interesting efforts to create molecules with conformational or functional properties of natural biopolymers have assembled the target materials using multistep reaction processes [1–4]. As an alternative synthetic approach, polymerization reactions offer one critical advantage: a complex material can be assembled in one process.

While the efficiency of polymerization processes makes them extremely attractive for neobiopolymer synthesis, the biological functions of the resulting materials will be difficult to dissect if the materials are highly heterogeneous. Consequently, developments in polymer chemistry that will enable the synthesis of homogeneous, highly functionalized materials are needed. Living polymerization processes offer opportunities to generate materials with these desirable attributes. A living polymerization is defined as a reaction in which the termination and chain transfer processes are slow relative to elongation. For reactions in which the rate of initiation exceeds that of propagation, materials of controlled molecular masses can be generated by varying the monomer-to-initiator ratios. Although many living polymerizations are known, the reaction conditions for most of these processes do not tolerate the dense display of polar functionality essential for the function of biological molecules. A number of research groups, therefore, have set out to discover and exploit polymerization methods that are compatible with the diversity of functional groups that characterizes biological systems.

Neobiopolymers have been generated by a number of different polymerization strategies. These include radical polymerization, cationic polymerization, ring-opening polymerization (ROP), and ring-opening metathesis polymerization (ROMP). The key mechanistic issues, advantages, and disadvantages of each method are outlined in this chapter. In addition, the recent, significant advances in ROMP for the synthesis of novel biopolymers will be discussed.

2
Examples of Methods Used to Generate Bioactive Polymers

2.1
Radical Polymerization

Free radical polymerization was the first method used to generate neobiopolymers (Fig. 1A) [5–8]. The polymerization reaction requires a radical initiator such as tetramethylenediamine (TEMED), ammonium peroxysulfate (APS), or 2,2'-azobisisobutyronitrile (AIBN) (Fig. 1B) [9]. Under optimized conditions, these compounds serve as initiators by forming the first radical species. The resulting radical can then react with a substrate, generally an electron deficient olefin such as acrylamide, which generates a new intermediate. From this radical species, the polymerization reaction propagates until the monomer is completely consumed. The chemistry of this process has important advantages, including a tolerance for monomers with polar, functional groups and the ability to conduct the reactions of such monomers in water. Large drawbacks of the radical approach exist, however, in that the polymers resulting from these processes often have high molecular weights and high polydispersities (a large distribution of polymer lengths).

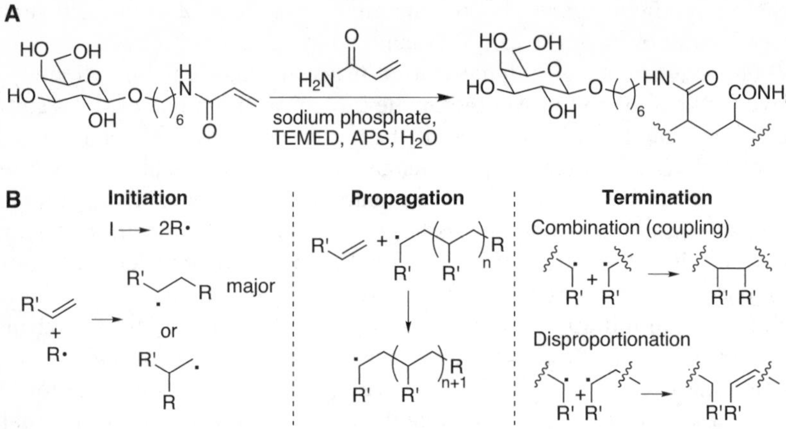

A

B

Fig. 1A,B. Application of radical polymerization to the synthesis of neobiopolymers. A Y.C. Lee's approach to galactose-substituted polyacrylamide gels. B The stages of radical polymerization

The product heterogeneity that results from radical polymerizations arises from inefficient initiation, chain transfer and termination events. Ideally, each initiating species, for example APS, would generate a reactive monomer, but initiation typically does not occur with the desired efficiency. Vinyl polymerizations, which are common for the generation of neoglycopolymers, typically have initiation efficiencies between 60% and 100% [9]. Under such conditions, control of polymer length is not feasible. This problem is difficult to solve. Developing molecules that undergo more rapid initiation is not enough; the intermediates generated are active enough to undergo chain transfer (addition of two growing chains) or termination. These alternative reaction processes add to the difficulty of generating materials of controlled lengths as well as complicating the synthesis of mixed copolymers (polymers incorporating different monomer templates). Despite these challenges, significant progress toward developing living free radical polymerizations has been made [10], and these methods have recently been applied to the synthesis of well-defined neoglycopolymers [11].

2.2
Cationic Polymerization

Another reaction that has been applied to the generation of highly functionalized polymers is cationic polymerization [12–15]. Catalysts for cationic polymerizations are aprotic acids, protic acids, or stable carbocation salts. In these processes, the catalyst generally reacts with a cocatalyst to form an active initiated species. Initiation takes place by protonation of the monomer (Fig. 2A). Monomers that possess cation stabilizing groups, such as electron rich olefins, are preferred as they more readily undergo the desired polymerization process

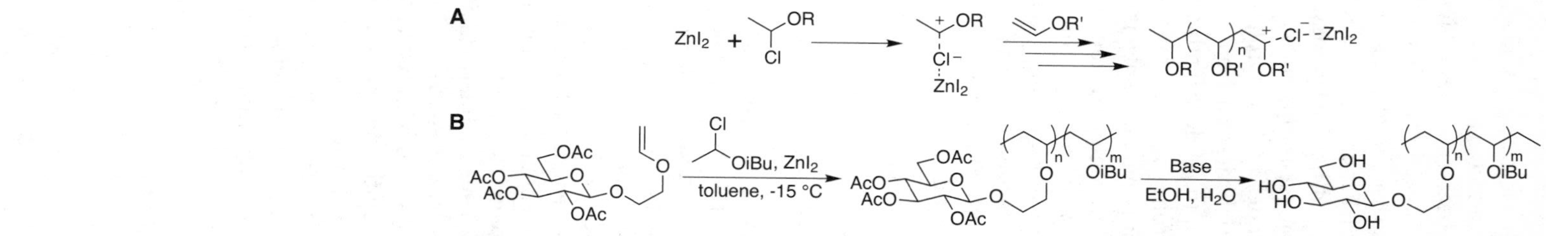

Fig. 2A,B. Application of cationic polymerization for neobiopolymer synthesis. **A** Mechanism of polymerization. **B** Minoda's use of Higashimura and Sawamoto's initiating system to generate a protected glucose-substituted polymer

with minimal side reactions. Because of the high reactivity of the carbocationic intermediates, an advantage of the electron-rich alkene monomers is that the polymerization reactions can be conducted at or below 0°C, thereby minimizing byproducts [16]. While there have been examples of living cationic polymerization and copolymerization, these reactions will not tolerate key functional groups found in biomolecules, such as multiple amides, nitrogen-containing heterocyclic rings and alcohols [16]. For example, carbohydrate-substituted polymers can be prepared by this method, but the hydroxyl groups must be protected (Fig. 2B) [13]. For unique functional groups, such as sulfate, found in biologically important natural carbohydrates and glycoproteins, no suitable protecting group strategy is available. Materials derived from peptides and nucleic acid components would also be difficult to produce by this strategy, given the diversity of nucleophilic functional groups relevant for their biological activities. Moreover, even for the synthesis of polymers displaying less functionalized saccharides, the need for protection can complicate the synthesis. For example, the efficiency of the protecting group removal for some masked polymers was shown to be dependent on the type of protecting group used [17, 18]. Another potential drawback is that the protecting group can interact with the catalyst system, as was suggested for the phthalimide group in a protected glucosamine monomer [13]. Therefore this method can be used to generate some bioactive polymers, but its scope is limited.

2.3
Ring-Opening Polymerization (ROP)

The generation of neobiopolymers through ring-opening polymerization (ROP) has recently been explored (Fig. 3) [19–21]. These polymers have unique structural features in that they share the amide backbone, and perhaps some of the conformational preferences of polypeptides and glycoproteins. The assembly of these materials can be initiated by basic or acidic catalysts, and propagation takes place through either anionic or cationic intermediates generated through ring-opening reactions [22]. Once generated, the intermediate, whether anionic or cationic, propagates by ring-opening of additional monomer units. Therefore, the design of the monomer template depends on whether the reaction is to be conducted through anionic or cationic intermediates (Fig. 3A,C). Carbohydrate-substituted materials have been prepared by both cationic and anionic ROP. For synthesis by the former method, a bicyclic, protected glucosamine-derived oxazoline was used as the initiating species for a cationic, living ROP (Fig. 3B), which afforded materials that terminate in a saccharide residue after deprotection [19]. This strategy was also used to form an interesting class of graft copolymers that displayed branches capped with glucosamine residues. In the case of anionic ROP, a saccharide-substituted α-amino acid N-carboxyanhydride (NCA) template was used to synthesize carbohydrate-containing polyamides (Fig. 3D) [20, 21]. Upon removal of the hydroxyl protecting groups, the resulting materials correspond to O-glycosylated polyserine. This structure has

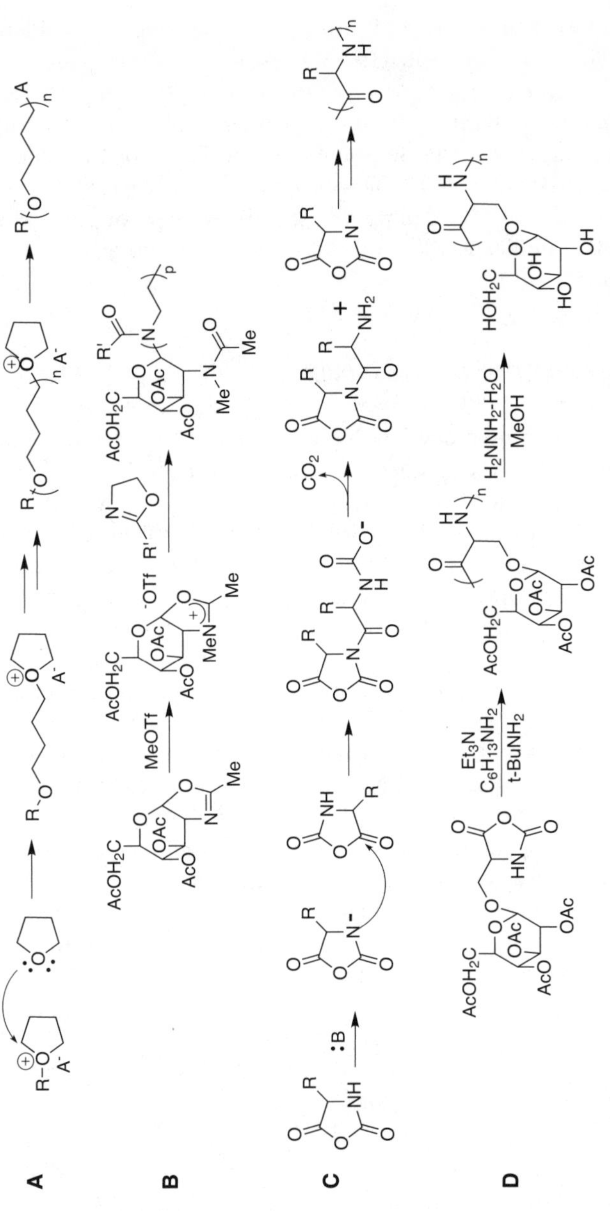

Fig. 3A–D. Use of ring-opening polymerization (ROP) for neobiopolymer synthesis: **A** General mechanism of cationic ROP. **B** Okada's use of cationic ROP to generate polymers with carbohydrate substituents at the terminus. **C** General mechanism for anionic ROP. **D** Okada's application of anionic ROP to generate carbohydrate-substituted polymers

many features of glycoproteins, yet it does not correspond to any identified in nature to date. The biological properties of these saccharide-substituted materials have not yet been characterized. As with cationic polymerization, the ionic intermediates in these polymerizations can be intercepted by reactive functional groups found in naturally occurring biopolymers. Thus, for the production of most biopolymer analogs by ROP, the reactions must be carried out using protected monomer units. The need to remove the protecting groups to generate the desired materials complicates the synthesis of many attractive bioactive materials by this method.

2.4
Ring-Opening Metathesis Polymerization (ROMP)

Recently, the ring-opening metathesis polymerization (ROMP) has emerged as an attractive approach to the synthesis of neobiopolymers. To date, four organometallic complexes, $RuCl_3$, $Cl_2(PCy_3)_2RuCHCHCHPh_2$, $Cl_2(PCy_3)_2RuCHPh$, and $Mo(CHCMe_2Ph)(N-2,6-i-Pr_2C_6H_3)(O-t-Bu)_2$, have been used to initiate neobiopolymer synthesis (Fig. 4). The advent of defined, functional group tolerant, metal alkylidene initiators has offered new opportunities for the creation of tailored materials with significant biological properties. A generalized mechanism for polymerization involves an initial coordination of the metal alkylidene ($RuCl_3$ is believed to react with the monomer to form a metal carbene, which is likely the active species) [23]. A [2+2] cycloaddition between the initiator and a monomer unit forms a metallocyclobutane which then undergoes a retro [2+2] ring opening (Fig. 4) [24]. A [2+2] cycloaddition of the resulting metal alkylidene with another monomer occurs, followed by ring-opening, and this sequence of events (propagation) continues until the monomer is fully consumed. The mechanistic pathway indicates that strained, cyclic alkenes will be highly reactive monomers, as is observed. Depending on the nature of the propagating intermediate, the growing chain can be terminated with an electron rich olefin if a ruthenium carbene is employed or an aldehyde in the case of the molybde-

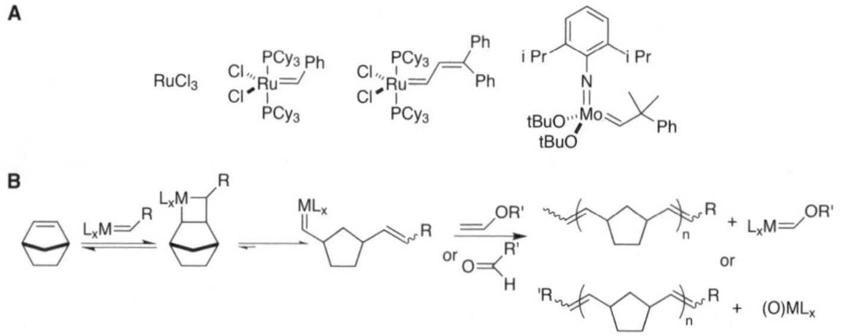

Fig. 4A,B. Ring-opening metathesis polymerization (ROMP): A Structures of organometallic initiators that have been used in ROMP to generate neobiopolymers. B General pathway for polymer synthesis using ROMP. Molybdenum-initiated reactions are typically capped with aldehydes and ruthenium-initiated with end ethers.

num alkylidene-catalyzed reaction. Although the overall mechanistic pathway is similar for polymerization with the molybdenum and ruthenium complexes, the initiators' tolerance of functional groups is different. For example, the ruthenium, but not the molybdenum, initiators tolerate unprotected hydroxyl groups and even sulfates, while the molybdenum initiator is more tolerant of amines [24, 25]. Importantly, however, the ROMP strategy provides new avenues for the production of interesting bioactive polymers. The remainder of this chapter will focus on the advances in ROMP technology in the generation of biopolymers and will include some of the significant biological problems that have been addressed.

3
Use of Organometallic Initiators to Make Bioactive Polymers

3.1
Use of RuCl₃ to Generate Carbohydrate-Substituted Polymers and Their Application to the Study of Multivalent Interactions

3.1.1
Synthesis of the Carbohydrate-Substituted Polymers

The first demonstration that ROMP could be used to generate bioactive materials focused on the preparation and properties of carbohydrate-substituted multidentate ligands [26]. Because carbohydrates are naturally presented on glycoprotein backbones or in glycolipid assemblies, saccharide-substituted polymers represent attractive tools with which to investigate protein–carbohydrate interactions. The polymerization of saccharide-substituted monomers was carried out using $RuCl_3$ as the initiating agent. Previous examples of ROMP suggested that characteristics such as control over polymer size and structure, the ability to generate copolymers, and tolerance of functional groups could be satisfactorily achieved, but demonstration that such materials could be used to explore and/or modulate biological recognition processes was lacking. Therefore, the immediate goal of these studies was to rapidly prepare carbohydrate-substituted materials to determine whether they could serve as ligands for carbohydrate-binding proteins. The monomer templates that were employed were composed of two carbohydrate recognition epitopes extending from a single 7-oxanorbornene scaffold (Fig. 5). Using either α-*O*- or *C*-linker units, saccharide residues were appended to the template through ester groups. The polymerization of these monomers in the presence of $RuCl_3$ in water proceeded in good yields (ca. 70–80%). Characterization of the molecular masses of the polymers was accomplished by gel filtration chromatography. Using dextran standards, the resulting polymers were determined to have relative molecular masses (M_r) in the range of 10^6. This route provided ready access to polymers displaying α-*O*- or *C*-linked glucose or mannose epitopes for biological testing.

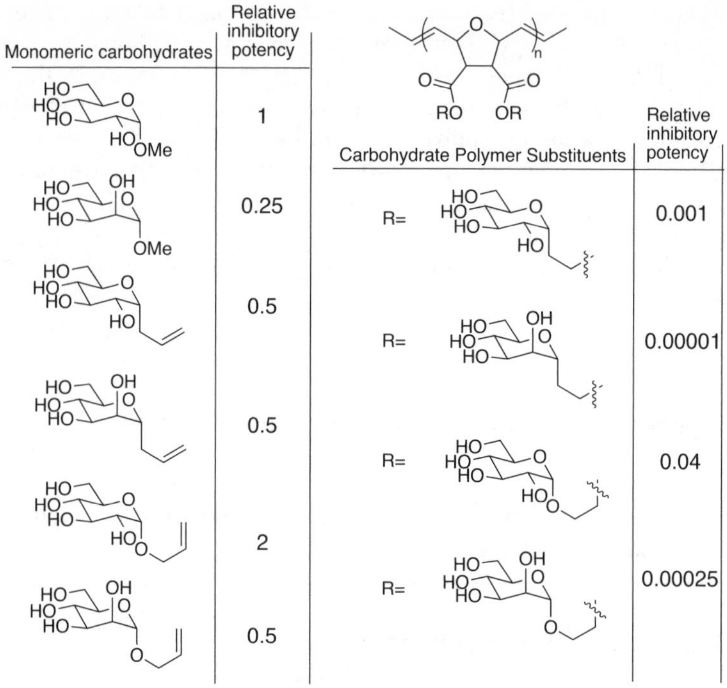

Fig. 5. Structure and biological activities of Kiessling's initial carbohydrate-substituted polymers generated by ROMP. Relative inhibitory potency: the saccharide residue concentration needed to inhibit the agglutination of red blood cells mediated by the protein concanavalin A

Because the accessibility of a saccharide epitope can dramatically affect its ability to be recognized by a protein, a second generation of polymers, designed to display saccharide residues at lower density, was generated [27]. These materials were composed of monomers substituted with a single carbohydrate epitope. The relevant monomer substrates could be easily synthesized as a mixture of diastereomers by treatment of the oxanorbornene anhydride with one equivalent of a protected carbohydrate-derived alcohol to open the anhydride, followed by addition of diazomethane to transform the liberated acid into a methyl ester. In addition to changes in the character of the monomer, modified polymerization conditions were investigated. Attempts to directly polymerize these substrates using ruthenium trichloride in water were unsuccessful affording low yields; attempts to optimize these conditions by addition of more initiator also were unsuccessful. A procedure to preform the active propagating species for use in subsequent polymerizations afforded good results [28]. A small amount of monomer was heated with the ruthenium trichloride for 12 h, and a portion of this mixture was subsequently added to the solution of monomer to be polymerized. Using this in situ initiator preparation, α-C-linked mannose

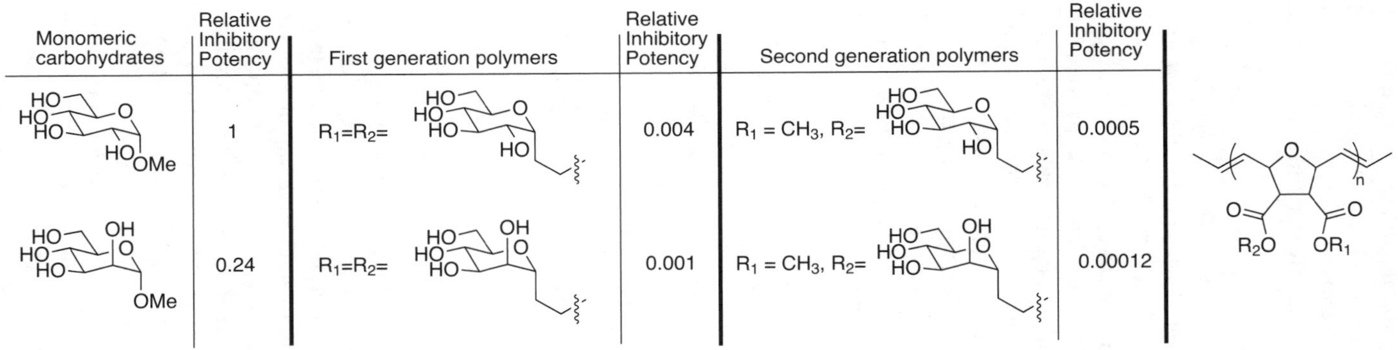

Fig. 6. Comparison of the biological activities of monovalent glucose and mannose derivatives, multivalent carbohydrate-substituted polymer with two saccharide epitopes per repeat unit, and the less sterically congested carbohydrate-substituted polymer with a single recognition element per repeat unit. All polymers were generated by ROMP using $RuCl_3$

and glucose polymers of the approximate same size and stereochemical composition as the first generation polymers (same percentage of trans double bonds) were generated (Fig. 6) for comparison. These materials constitute a series of compounds that was used to investigate structure-function relationships in protein–carbohydrate interactions.

3.1.2
Bioactive Materials from ROMP: Exploring Multivalent Interactions

Multidentate displays of saccharide epitopes can be used to elucidate the mechanistic features that contribute to protein–carbohydrate interactions, and the first and second generation polymers were designed for this purpose. A critical issue that these materials were designed to address is how low affinity binding, such as that which characterizes many protein–saccharide interactions, can give rise to affinity and specificity in vivo. Specifically, the binding affinity of a protein to a single carbohydrate residue is often weak ($K_d \approx 10^{-3}$ M) and many carbohydrate-binding proteins (lectins) can recognize related, but distinct, saccharide epitopes [29]. Carbohydrates in physiological settings, however, are rarely found as isolated saccharide units. These entities are more commonly presented in arrays in which multiple copies of saccharide residues are displayed, as is observed for glycolipid assemblies and glycoproteins. The enhanced interactions available to a multivalent display may provide the key to understanding how such molecules can mount the necessary avidity and selectivity for protein–carbohydrate recognition processes in vivo. The exact mechanisms by which different multivalent displays achieve this objective are not known, however, and these may differ for particular receptor-ligand complexation events. Some of the possibilities for binding enhancement through multivalency include: binding of a multivalent ligand to a multivalent receptor (the chelate effect), clustering of receptors by a multivalent ligand, subsite binding, and increased local concentration (Fig. 7). Access to materials which present multiple copies of specific carbohydrate epitopes will facilitate elucidation of the underlying forces governing these recognition events.

Concanavalin A (ConA), a tetrameric plant lectin that binds both glucose and mannose residues, is an excellent model protein with which to probe the importance of multivalent displays in protein-carbohydrate interactions [30]. The binding of a wide range of mono- and multivalent saccharide derivatives to ConA has been investigated, and a tremendous amount of structural and biophysical data has been amassed. The features of ConA binding to *O*-glycosylated α-mannose and glucose residues are well characterized: ConA exhibits a fourfold preference for α-mannose over α-glucose. Investigations of the corresponding α-*C*-linked carbohydrates reveal that there is little difference in the binding free energies of α-*C*-propyl mannoside and α-*C*-propyl glucoside, although a subtle preference (1.5-fold) for the mannose derivative may exist [31]. With monomeric and polymeric *C*- and *O*-glycosides of mannose and glucose, the affinity of ConA for the multivalent versus monovalent epitopes could be deter-

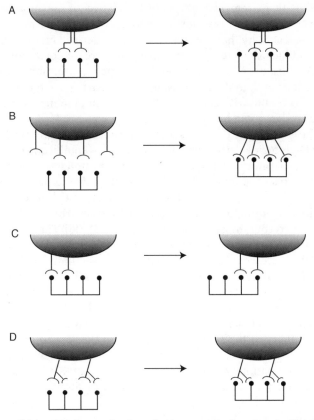

Fig. 7A–D. Possible molecular mechanisms for increases in functional affinities for multivalent ligands. **A** Interaction of a multivalent ligand with a multivalent receptor (chelate effect). **B** Clustering of receptors by a multivalent ligand. **C** Increased local concentration (slow off-rates by statistical effect). **D** Subsite binding (occupation of secondary binding sites on a single receptor)

mined, as well as the specificity for binding one multidentate ligand over the related compound.

The activities of the monovalent and multivalent ligands to interact with ConA could be assessed by evaluating their abilities to inhibit ConA facilitated agglutination of red blood cells. Because ConA is a tetramer at neutral pH, it can bind glycoproteins on opposing red blood cell surfaces thereby acting like a molecular glue that holds cells together. To assess and compare the efficacy of the various compounds, the relative inhibition values were calculated on a saccharide residue basis by determining the molar concentration of saccharide residues required to block cell agglutination.

In the case of first generation polymers, which display two saccharide epitopes for each monomer unit, the α-C-linked carbohydrate-substituted polymers bound with extremely high avidities relative to the monovalent compounds. Specifically, the C-mannoside polymer was 105-fold more effective at inhibiting ConA activity than was the C-mannose-derived monomer. Moreover, the multivalent ligands were more selective than were the corresponding monovalent compounds: the C-mannose-derivatized polymer inhibited at a 100-fold lower concentration than its C-glucoside counterpart. Since the monomeric α-C-linked sugars bound with equal affinities, this result highlights the greater selectivity multivalency can impart.

Overall the C-glycoside polymers were more effective inhibitors than the corresponding O-glycoside derivatives, but both showed significant avidity increases relative to the monovalent glycosides. It should be noted that the less polar C-glycosides may result in increases in ConA binding due to the hydrophobic effect. With regard to the specificity of interactions, the α-O-linked mannose polymer inhibited agglutination at a 160-fold lower saccharide residue concentration than the corresponding glucose polymer (Fig. 5) [26]. The increase in selectivity for mannose over glucose displayed by the O-glycoside polymers is consistent with the increase in affinity differences between individual mannose and glucose residues (4-fold for the O-glycosides versus 1.5-fold for the C-glycosides). The second generation polymers showed similar trends; however, these materials were approximately tenfold more potent than for the first generation polymers (Fig. 6) [27]. This difference may be explained by considering the greater accessibility of the saccharide residues in the second generation polymers; these epitopes are less sterically encumbered by adjacent saccharide residues. These investigations were the first to establish that ROMP could be used to prepare biologically active materials and that these materials could be used to illuminate fundamental biological processes. The ability to create materials of different molecular masses, to generate block copolymers, and to specifically end-label the polymers for immobilization or detection, would facilitate a range of biological studies. The possibilities provide impetus, from a new direction, for advances in polymerization technologies and their applications.

3.2
Use of Cl$_2$(PR$_3$)$_2$RuCHCHCHCPh$_2$ (R=Ph or Cy) to Make Carbohydrate-Substituted Polymers

3.2.1
Synthesis of Polymers Substituted with Protected Carbohydrate Residues

The first report of a neobiopolymer made with a defined organometallic complex, Cl$_2$(PR$_3$)$_2$Ru=CHCH=CPh$_2$ (R=Ph or Cy), was contributed by Fraser and Grubbs [18]. The neobiopolymers produced had a norbornene backbone with pendant, protected glucosamine residues (Fig. 8). Using monomers equipped with carbohydrate residues protected with acetate, triethylsilyl, or benzyl groups, the polymerizations were conducted in a range of different solvents and terminated by addition of ethyl vinyl ether. These polymerizations had some of the characteristics of living processes. For example, gel permeation chromatog-

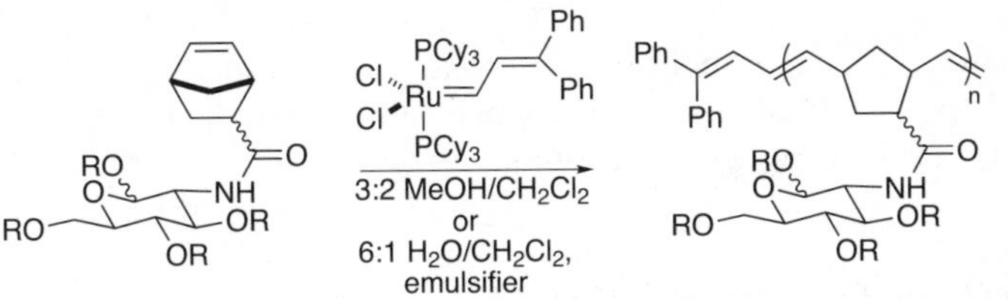

Fig. 8. First synthesis of carbohydrate-substituted polymers by ROMP using a defined organometallic initiator. Grubbs' synthesis of glucosamine-substituted polymers

raphy (GPC) revealed that the polydispersities for several of these materials were low (ca. 1.1–1.2). In addition, diblock copolymers could be generated from this process, but it was not established that these conditions truly resulted in living polymerization of the saccharide-substituted monomers. The polymerization of protected carbohydrate-modified monomers was successful. The different protecting groups, however, resulted in significant alterations in the rates of reaction of the various monomers, with acetate protected monomers reacting quickly (<5 min) and silyl ether protected monomers very slowly (2–3 days). In addition to complexities resulting from differences in the rate of polymerization, the efficiencies of the different protecting group removal were variable. Although the silyl ether protecting groups were easily cleaved with tetrabutylammoniumfluoride in THF, attempts to remove other blocking groups resulted in incomplete reactions and the formation of side products.

3.2.2
Synthesis of a Polymer Possessing Unprotected Carbohydrate Residues

An important observation for the design and synthesis of neobiopolymers is that an unprotected glucose-substituted monomer could be polymerized using defined, metallocarbene ROMP initiators [18]. The assembly of unprotected monomers circumvents the problems of protecting group installation and removal and perhaps also that of differential rates of monomer reaction; moreover, the opportunities increase for engineering materials with desirable properties with the defined, well-characterized initiators. In these studies, the solubilities of the hydrophilic monomer and the hydrophobic metal complex are so divergent that the reaction conditions used for the assembly of protected monomers were unsuccessful. The problem was solved by the application of aqueous emulsion conditions, in which a cationic detergent was used to promote reaction at the aqueous-organic interface. The reactions were conducted in a dichloromethane-water mixture in the presence of the surfactant dodecyltrimethylammonium bromide (DTAB), and the reactions were terminated by the addition of ethyl vinyl ether. Under these conditions, the polymerization reaction proceeded in high yields (99%) to afford glucose-substituted polymers. The characterization of these materials is complex, as GPC analysis is generally conducted in organic solvents. Nevertheless, GPC analysis of polymers soluble in organic solutions, but generated under the same conditions, indicated that polymers of low polydispersity (PDI≤1.1) were formed [32]. These data suggest that these conditions may be conducive to a living polymerization.

3.3
Use of Mo(CHCMe₂Ph)(N-2,6-*i*-Pr₂C₆H₃)(O-*t*-Bu)₂ to Generate Carbohydrate-Substituted Polymers

Shortly following initial reports of neobiopolymer formation using ruthenium carbene complexes, the Schrock group reported the synthesis of "sugar-coated" polymers using a molybdenum complex, Mo(CHCMe₂Ph)(N-2,6-*i*-Pr₂C₆H₃)(O-*t*-Bu)₂ (Fig. 9) [17]. These studies were the first to demonstrate that a living polymerization can be achieved with monomers bearing protected saccharide residues. Although molybdenum alkylidenes are not as tolerant of functional groups, these reagents have several desirable features that can be exploited in the preparation of neobiopolymers. Specifically, these initiators can be used to generate stereodefined materials because they allow control over the *cis/trans* ratio of polymer backbone alkenes and polymer tacticity [33–37]. Moreover, elongation can be terminated conveniently with functionalized aldehydes, which can give rise to polymers substituted with specific reporter end groups [38].

The polymerization of several different protected monomers was executed with the molybdenum initiator. Two types of norbornene templates were generated as mixtures of diastereomers: one class possessed a single saccharide substituent attached through either an *endo* or *exo* ester group, and the other was equipped with two carbohydrate residues appended to the bicyclic scaffold

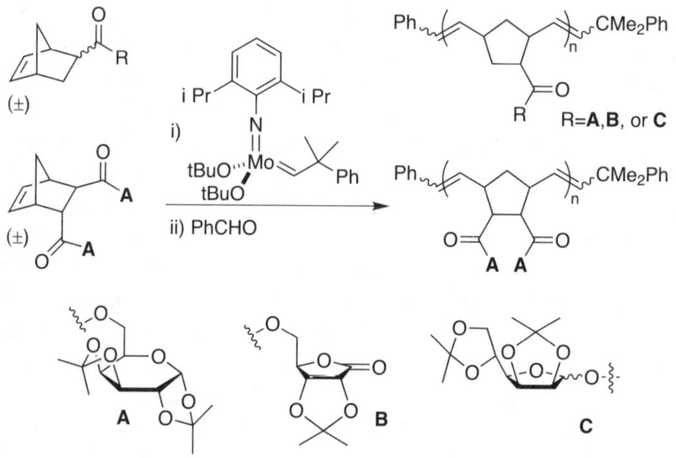

Fig. 9. Living polymerization to produce carbohydrate-substituted polymers by Schrock utilizing a defined molybdenum initiator

through *trans* oriented ester groups. The saccharides used were derived from galactose, mannose, and ribonic-γ-lactose, and in each, the hydroxyl groups were
masked through cyclic acetal formation. The reactions were conducted in toluene with $Mo(CHCMe_2Ph)(N-2,6-i-Pr_2C_6H_3)(O-t-Bu)_2$, the most functional
group tolerant molybdenum initiator, and quenched by addition of benzaldehyde. The product polymers were obtained in high yields (92–99%) and were of
low polydispersity (1.02–1.25), as determined by GPC. In addition to these results, which suggest the polymerization is living, several multiblock polymers
were also generated. Therefore, Schrock demonstrated that carbohydrate substituted polymers could be assembled through living ROMP. Unfortunately, attempts to remove the saccharide protecting groups were not always successful.
While treatment of two of the polymers with trifluoroacetic acid (TFA) and water resulted in the desired products of acetal cleavage, these conditions failed
with the other two polymers. The NMR data from the two unprotected polymers
are consistent both with protecting group cleavage and retention of the saccharide epitopes; however, the spectroscopic studies cannot rule out the possibility
that some side reactions have taken place. Moreover, some biopolymer functional groups, such as heterocyclic bases, phosphoryl groups, and sulfates,
would be incompatible with these reaction conditions. Although the molybdenum initiator used in these initial studies afforded mixtures of backbone stereoisomers, such catalysts may provide access to stereochemically defined materials. A concern, however, is that the scope of saccharide residues that can be incorporated using known molybdenum alkylidene complexes may be limited.

3.4
Use of $Mo(CHCMe_2Ph)(N-2,6-i-Pr_2C_6H_3)(O-t-Bu)_2$ to Generate
Amino Acid-Derived Polymers

Concurrent with early studies using ROMP to generate saccharide substituted
polymers, Gibson and coworkers were exploring the utility of ROMP for the production of amino acid-derived polymers (Fig. 10) [39]. Using the functional
group tolerant molybdenum initiator, polymers substituted with alanine residues were generated. The monomer units used were derived from attachment of
alanine methyl ester to an *exo-* and *endo*-norbornene imide template. The template was generated by reaction of the alanine amino group with the correspond-

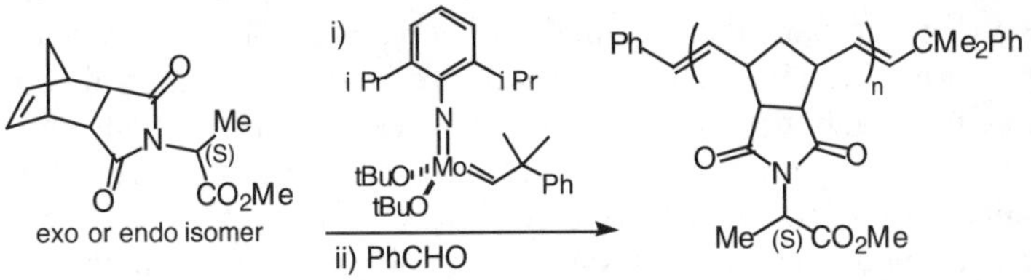

Fig. 10. Gibson's synthesis of amino acid-derived polymers using a defined molybdenum
initiator

ing norbornene anhydride derivative, a process that occurred without racemization of the amino acid [40]. The polymers were generated by treatment of the monomer with the molybdenum initiator in benzene. The propagation reactions were terminated with benzaldehyde, and the polymers were isolated by precipitation with hexanes. Characterization of the polymers by GPC revealed a narrow molecular mass distribution, with PDI values between 1.13 and 1.29. In addition, the product polymers were found to contain mostly *trans* double bonds (77–91%). By employing the symmetrical monomer with the molybdenum carbene, highly stereochemically regular polymers could be synthesized. Thus polymerization demonstrates that simple amino acid residues can be successfully incorporated into polymers using ROMP, and the results suggest that materials that have little stereochemical heterogeneity can be synthesized [41]. The authors envision that this approach can give rise to materials in which the amino acid side chains influence polymer structure, presumably generating materials with unique architectures.

3.5
Use of $Cl_2(PCy_3)_2RuCHPh$ for the Synthesis of Thymine-Substituted Polymers as Potential Nucleic Acid Analogs

The design of DNA and RNA analogs with non-natural backbones is an area that has prompted considerable studies. The goal has been to generate molecules that conserve the recognition properties of oligonucleotides, to target endogenous nucleic acids for example, while incorporating desirable properties not possessed by natural oligonucleotides, such as increased stability and the ability to rapidly cross membranes. Phosphorothioate-substituted oligonucleotides, in which a non-bridging phosphoryl oxygen group has been replaced with sulfur, are close analogs of oligonucleotides that share their recognition properties but have increased stability toward degradation by nucleases [42]. Another class of backbone-modified analogs is protein nucleic acids (PNAs), which are neutral oligonucleotide mimics. These compounds can selectively bind RNA and DNA sequences, and they do so with much higher affinities than oligonucleotides because they do not experience the repulsive electrostatic forces that must be overcome in complexation. Gibson and coworkers explored the synthesis of a distinctly different type of neutral oligonucleotide analog, which they envisioned could be generated by ROMP.)

Gibson and coworkers described the synthesis of thymine-substituted polymers as a first step towards the generation of nucleic acid analogs using ROMP [43]. The norbornene imide template used previously in the assembly of amino acid-substituted polymers was employed, and a short linker used to connect the template and the thymine residue (Fig. 11). The major change from the Gibson group's previous work was the use of a ruthenium, rather than a molybdenum, alkylidene initiator, a choice that was presumably driven by the higher functional group tolerance exhibited by the ruthenium complex. The reaction products, which were generated by polymerization in THF and termination by the addi-

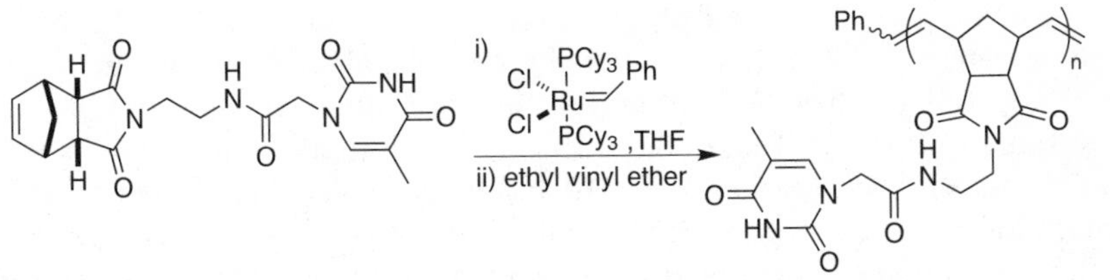

Fig. 11. Gibson and coworkers synthesis of a polymer displaying the nucleobase thymine using ROMP initiated by Grubbs' ruthenium complex

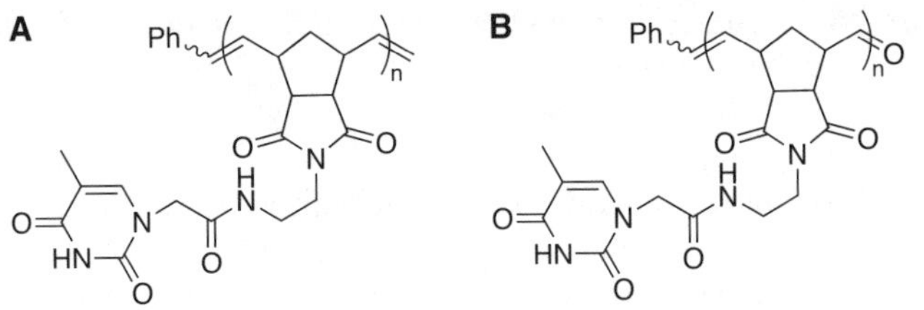

Fig. 12. Analysis of the thymine-substituted polymers by MALDI-TOF mass spectrometry reveals that the polymers can be capped with two different groups

tion of ethyl vinyl ether, were characterized by proton nuclear magnetic resonance (NMR) spectroscopy and matrix-assisted laser desorption ionization time of flight (MALDI-TOF) mass spectrometry. The analytical characterization of the thymine-substituted materials provided insight into the polymerization process. Using proton NMR spectroscopy, the reaction could be monitored by following the transformation of the initiating ruthenium species into a propagating species. MALDI-TOF mass spectrometry yielded additional information about the reaction features, and this study provided an excellent example of the utility of this technique in polymer analysis. When the polymerization was conducted with three equivalents of thymine-substituted monomer, the resulting product had a low PDI, 1.07. The observed M_n of 2500, however, was higher than the calculated M_n of 1221. This indicates that propagation is faster than initiation for this monomer under these conditions. Rates of initiation and propagation have been shown to be dependent on many reaction conditions, such as substrate structure, solvent conditions, and monomer-to-initiator ratios [18]. Another notable finding is that the conditions used for termination influenced the structure of the final product. When ethyl vinyl ether was used to terminate the polymerization, the major signals from the MALDI analysis could be attributed to polymers end-capped with a methylidene group. In contrast, exposure of the polymerization reaction to oxygen resulted in aldehyde capping groups as detected by MALDI spectrometry. This capping process for ruthenium carbenes had not previously been noted in the literature (Fig. 12).

The investigations directed at the synthesis of thymine-substituted polymers demonstrate that the type of functional groups displayed by nucleic acid bases are compatible with ROMP. Moreover, the application of MALDI-TOF mass spectrometry to the analysis of these polymers adds to the battery of tools available for the characterization of ROMP and its products. The utility of this approach for the creation of molecules with the desired biological properties, however, is still undetermined. It is unknown whether these thymine-substituted polymers can hybridize with nucleic acids. Moreover, ROMP does not provide a simple solution to the controlled synthesis of materials that display specific sequences composed of all five common nucleic acid bases. Nevertheless, the demonstration that metathesis reactions can be conducted with such substrates suggests that perhaps neobiopolymers that function as nucleic acid analogs can be synthesized by such processes.

3.6
Use of Cl$_2$(PCy$_3$)$_2$RuCHPh to Generate Multivalent Penicillin Derivatives

Molecules that may have biological activity, but are not designed to mimic naturally occurring biopolymers, also have been synthesized [44]. One design entailed appending a bioactive agent, such as penicillin, to a poly(norbornene) backbone. Baigin et al. outline several potential uses for such polymers: new bandage materials, drug delivery systems, or affinity column materials. In this example, polymers presenting multiple copies of 6-N-acylamino penicillanic acid were assembled using the Grubbs' ruthenium carbene initiator (Fig. 13). The polymer products were analyzed by proton NMR spectroscopy, and MALDI-TOF and electrospray mass spectrometry. In contrast to the excellent results obtained in the analysis of the thymine-substituted polymers, good quality MALDI-TOF spectra could not be obtained. Consequently, other mass spectrometry methods for analyzing large molecules were employed. A material generated with ten equivalents of monomer relative to initiator was examined by negative ion electrospray mass spectrometry. The spectra contained major peaks corresponding to polymers between 8 and 17 monomer units long. The average polymer length, which was estimated by comparing the integration of the NMR signals arising from the hydrogen atom attached to the ruthenium

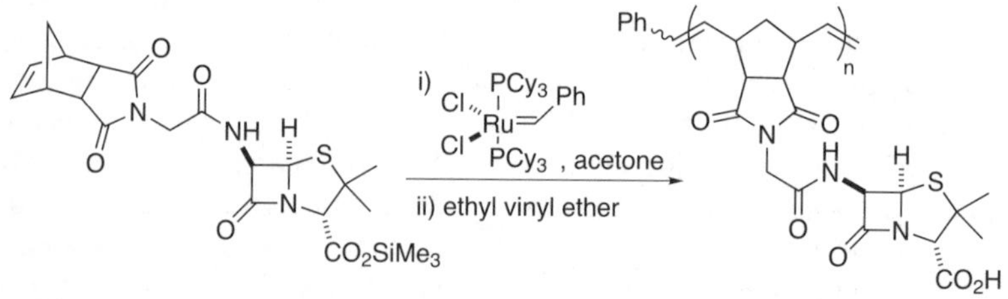

Fig. 13. The generation of multivalent penicillin derivatives by ROMP. It is proposed that such materials may function as drug delivery agents

carbene terminus to those of the backbone alkenes, was 24. Given the sensitivity of β-lactams and their potential for decomposition through a number of different mechanisms, this investigation provides additional examples of the diversity of functional groups tolerated by the ruthenium carbene complexes. The utility of these agents as drug delivery vehicles has not yet been described.

3.7
Use of $Cl_2(PCy_3)_2RuCHPh$ to Generate Carbohydrate-Substituted Polymers

3.7.1
Generating Carbohydrate Polymers of Defined Length

Together, the examples from the Grubbs, Schrock, and Gibson research groups provide ample evidence that polymers containing the critical functional groups used by biomolecules in physiological settings can be synthesized through living ROMP. These results are especially significant given the findings that materials created by ROMP can have potent biological activities [26, 27, 45–47]. The ability to control relevant aspects of polymer structure provides new tools for investigating the relationship between structure and function in biological systems. Multivalent protein–carbohydrate interactions, which are difficult to study using many traditional biochemical methods, but which are widespread in physiological settings, provide an excellent subject for study using materials created by ROMP.

An important issue in multivalent binding is to examine how the number of recognition elements affects the activities of a series of multidentate ligands. Living ROMP can be used to address this issue, as materials of defined molecular masses can be assembled and their biological activities tested. Specifically, a living polymerization in which initiation rates are faster than propagation provides a means to control polymer length by adjusting the monomer-to-initiator (m/i) ratio. To take advantage of this potential, Kiessling and coworkers used living ROMP to synthesize a series of mannose-substituted polymers of different average lengths.

The target mannose-substituted materials were constructed using the defined ruthenium alkylidene, $Cl_2(PCy_3)_2RuCHPh$, but the monomer structure as well as the reaction conditions needed to be carefully crafted to generate the desired products. Only the most reactive monomer templates resulted in controlled assembly of the polymers to afford materials of predictable lengths. The most reactive saccharide-substituted monomer was found to be a tricyclic norbornene imide template displaying a single α-linked mannose residue, a monomer design that is similar to that employed by the Gibson group. Less electron-rich alkenes, such as 7-oxanorbornene derivatives, or less strained alkenes, such as norbornenes bearing an exocyclic amide linkage, exhibited decreased reactivity and engaged in premature termination processes. The reaction conditions significantly affected the assembly process, with different monomer-to-initiator ratios exhibiting different requirements. To synthesize polymers of varying

A

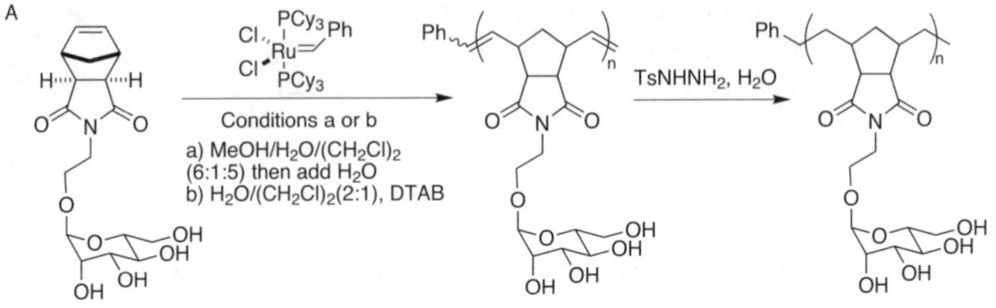

B

DP	Relative Potency
10	100
25	300
52	2,000
143	2,500
3,000	2,000

Fig. 14. A,B. A living polymerization of mannose-substituted norbornene derivatives can be used to produce materials of defined lengths for biological testing. **B** The relationship between polymer length and biological activity was explored

lengths, homogeneous conditions were employed for m/i ratios less than 100/1, but emulsion conditions were used for m/i ratios greater than 100/1 (Fig. 14A) [48]. The resulting materials, with their alkene-containing backbones, also provide an opportunity to explore the importance of conformational entropy, manifested in backbone flexibility, in their biological activities. The flexibility of the polymer could be increased by reducing the backbone alkene groups with diimide. Thus, the living polymerization provides a new set of tools for investigating biological recognition events.

3.7.2
Relationship Between Polymer Length and Binding Enhancement

It was envisioned that the different length mannose-substituted polymers could provide additional insight into multivalent binding events, such as those which are required for concanavalin A (ConA)-mediated agglutination of red blood cells. The Kiessling group's initial studies revealed that multidentate mannose-substituted polymers were potent inhibitors of the cell agglutination process. The underlying mechanism by which these multidentate ligands function, however, was unknown.

Since the cell agglutinating form of ConA is a tetramer, a simple model for the potent activity of the polymers would be that the saccharide recognition elements presented by these agents can simultaneously bind two or more sites on the protein. A structural study of ConA reveals that adjacent saccharide binding sites within the tetramer are separated by 65 Å, suggesting that the most potent polymers should be capable of spanning this distance. To determine whether saccharides attached to the polymer backbone could simultaneously occupy two sites, the activities of polymers of various lengths were assessed. As was expected from previous results, the multivalent compounds were more effective inhibitors of the agglutination process than the corresponding monomeric mannose derivative. Most significantly, however, the inhibitory potency (determined on a saccharide residue basis) of each polymer increased with increasing polymer length. Specifically, an exponential increase in potency was seen for linear polymer length increases up to a degree of polymerization (DP) of 50 (Fig. 14B). Interestingly, molecular mechanics modeling of the polymer backbone indicates that at a DP of 50 or above almost all polymers in solution are of sufficient length to bridge two saccharide binding sites on ConA. Above a DP of 50, the potency remained high, but did not dramatically increase. Similar effects were observed with the reduced polymers, indicating that increasing the conformational entropy of the polymer backbone has little effect on the activity of the polymers.

The activities of the polymers can be rationalized by considering the chelate effect as well as other statistical effects. The chelate effect is a manifestation of an increase in the apparent activities of binding for multidentate ligands and receptors: once a recognition element occupies the first binding site, the probability that a tethered recognition element will interact with a second ligand binding site is increased [49–51]. Thus, short mannose-substituted polymers will not be able to simultaneously occupy two saccharide binding sites, but as the length of the polymer population is increased, marked enhancements in inhibitory potencies should be observed (Fig. 7). The exponential augmentation in activities seen as the DP is varied from 10 to 50 units provides further support that the multivalent carbohydrate ligands are functioning through the chelate effect. As more polymers occupy two binding sites, their apparent affinity for ConA increases exponentially. Once the polymers are of sufficient length that both binding sites are spanned, however, a decrease in potency on a per saccharide residue basis might be expected. This decrease is not observed. Thus, it appears that the addition of residues still contributes to polymer activities (Fig. 14B). This contribution may be due to the statistical effect, where an increased local concentration of the a-linked mannose recognition element results in a decrease in off-rate for polymer dissociation (Fig. 7). Secondary site binding could also contribute to polymer activity (Fig. 7). The investigations with saccharide-substituted polymers (neoglycopolymers) provide insight into the molecular mechanisms by which multivalent ligands function in a physiological setting.

3.8
Use of Cl₂(PCy₃)₂RuCHPh to Generate Mixed Copolymers and Block Copolymers

3.8.1
Synthesis of Mixed Saccharide-Substituted Copolymers

The relationship between polymer length and bioactivity for the materials designed to target ConA suggested that the most important mechanism by which multidentate materials can achieve high apparent binding activities is through the chelate effect. These studies also indicated, however, that not all of the saccharide epitopes were contributing equally to binding. Because only some of the carbohydrate substituents are interacting with the protein and because the activities of the neoglycopolymers are assessed on a saccharide residue basis, it might be expected that materials containing fewer saccharide recognition epitopes located at critical positions would be more potent when evaluated on a saccharide residue basis. To further explore the relationship between polymer activity and number of recognition elements, Kanai et al. developed an approach to the synthesis of mixed copolymers and block copolymers, all of which contained carbohydrate recognition elements [52].

Mixed copolymers were synthesized to test the hypothesis that only a subset of recognition elements was needed to achieve high affinity binding (Fig. 15). Although a number of "spacer" residues could be employed, it was postulated that using a monomer unit of similar structure and solubility would facilitate the synthesis of defined polymers. Therefore, in addition to the norbornene imide template substituted with a mannose recognition element, a structurally related galactose-bearing monomer was used as the spacer element. The tricyclic norbornene template was retained in the spacer unit to preserve the reactivity of the monomer, but galactose, which has a solubility similar to that of mannose but which is not bound by ConA, was attached as the substituent. Because both monomers appeared to have the same reactivity, polymers composed of the two units should contain a statistical mixture of mannose and galactose residues, which is dependent on the ratio of mannose and galactose monomers in the reaction mixture. Using this strategy, two sets of polymers of average DPs of 10

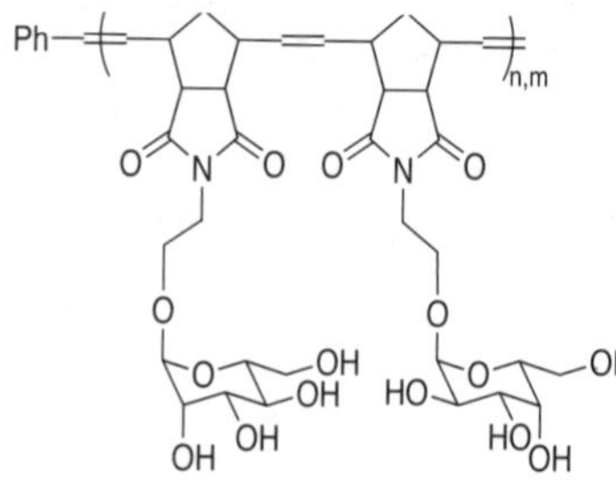

Fig. 15. Mixed copolymers were generated to examine the effect of changing the recognition epitope density for a biologically active series of carbohydrate-substituted polymers. Note: n and m represent ratios of mannose and galactose residues, respectively.

and 100 were assembled. For each set, several materials composed of different galactose and mannose ratios were synthesized. These mixed neoglycopolymers should provide insight into the importance of saccharide residue density on bioactivity.

3.8.2
Synthesis of Saccharide-Substituted Block Copolymers: Mimics of Glycoprotein Mucins

The mixed, random copolymers could be used to examine the relationship between saccharide epitope density and bioactivity, but block copolymers might be used to investigate the spacing between saccharide binding sites. Moreover, block copolymers could be useful for drug targeting strategies or for creating artificial vaccines. Interestingly, block copolymers share many features with a class of glycoproteins, termed mucins, which display clusters of O-linked saccharides separated by unglycosylated peptide segments. Kanai et al. sought to generate polymers that could mimic the clustered glycoside displays of mucins using triblock copolymers with mannose residues displayed at each end of the polymer (Fig. 16) [52]. The separating units could contain non-binding carbohydrates (e.g., galactose residues as used previously, Fig. 16A) or a simple spacer (e.g., a diol, Fig. 16B). In order to spectroscopically monitor and differentiate the incorporation of the first mannose sequence separately from the second, it was envisioned that 7-oxanorbornene imide-linked mannose be used at one end of the polymer and a norbornene imide-substituted mannose at the other terminus. Unfortunately, initial attempts at polymerization by sequential addition of oxanorbornene imide-substituted mannose, norbornene imide-derivatized galac-

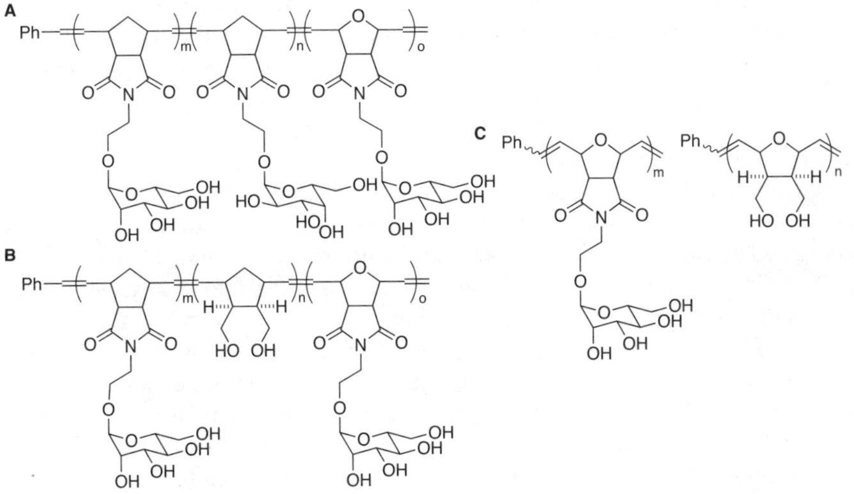

Fig. 16. A ruthenium initiator was used to create block copolymers to investigate the relationship between biological activity and the spacing of carbohydrates on the polymer backbone

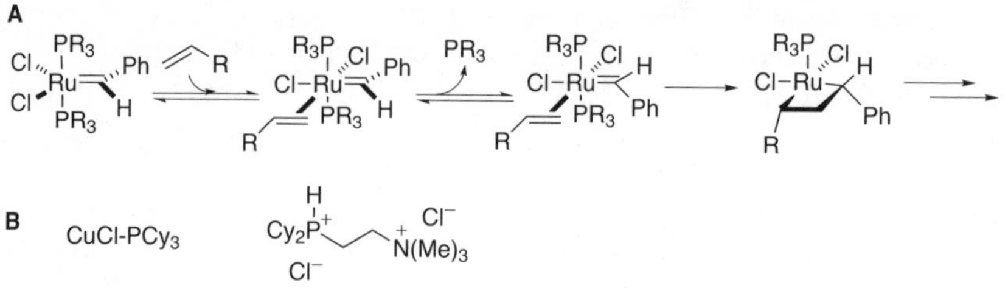

A

B   CuCl-PCy₃

Fig. 17. Mechanistic studies of alkene metathesis provided methods to increase the rates of reaction. Copper (I) chloride and hydrochloric acid have been used to encourage dissociation of the phosphine ligands

tose, and norbornene imide-appended mannose were unsuccessful, leading to a polymer approximately two times the expected length.

The problem was solved when new mechanistic insights into the process of alkene metathesis by the ruthenium initiator were revealed by the Grubbs' group. Their investigations indicated that dissociation of a phosphine from the ruthenium complex facilitated the metathesis reaction (Fig. 17A) [53]. The addition of phosphine scavengers, such as copper(I) chloride or hydrochloric acid, afforded large increases in the rates of metathesis and resulted in higher yields of metathesis products. The scavengers presumably act by trapping the released phosphine (Fig. 17B), either through complexation or protonation (Fig. 17B) [54].

For the synthesis of carbohydrate-substituted block copolymers, it might be expected that the addition of acid to the polymerization reactions would result in a rate increase. Indeed, the ROMP of saccharide-modified monomers, when conducted in the presence of *para*-toluene sulfonic acid under emulsion conditions, successfully yielded block copolymers [52]. A key to the success of these reactions was the isolation of the initiated species, which resulted in its separation from the dissociated phosphine. The initiated ruthenium complex was isolated by starting the polymerization in acidic organic solution, from which the reactive species precipitated. The solvent was removed, and the reactive species was washed with additional degassed solvent. The polymerization was completed under emulsion conditions (in water and DTAB), and additional blocks were generated by the sequential addition of the different monomers. This method of polymerization was successful for both the mannose/galactose polymer and for the mannose polymer with the intervening diol sequence (Fig. 16A,B).

As had been observed in the synthesis of carbohydrate-substituted polymers of different lengths, the reactivity of the monomers was an important parameter in generating the triblock polymers. If the mannose-substituted 7-oxanorbornene derivative was first polymerized, followed by the galactose-derivatized norbornene and the mannose-substituted norbornene monomers, two distinct sets of products were observed. These were identified by modification of the resulting polymers by acetylation, and analysis of the products by GPC. With this protocol, it was found that the product was composed of short polymers (DP=

15) and long polymers (DP=56), indicating that the reaction of the 7-oxanor-bornene derivative afforded significant amounts of terminated products. Attempts to use the emulsion polymerization conditions with another spacer monomer, the norbornene diol derivative, resulted in the isolation of two separate polymers, one which was due to the polymerization of the diol and the other from the mannose-substituted monomer (Fig. 16C). With monomers of different solubilities, the rates of phase transfer complicate the polymerization process. Homogeneous triblock polymers could be generated by altering the sequence of monomer additions. The most reactive mannose-substituted monomer, the tricyclic norbornene imide derivative [55], was polymerized first to minimize termination events. Treatment of the resulting intermediate with either of the spacer monomers and then with the 7-oxanorbornene-appended mannose derivative proceeded smoothly. The random and block copolymers can be used to develop additional structure and function relationships that will further illuminate multivalent binding events. These synthetic studies, however, are significant in the absence of biological data. They illustrate some of the methods that can be employed to overcome the obstacles encountered in the synthesis of water-soluble polymers by living ROMP.

3.9
$Cl_2(PCy_3)_2RuCHPh$ Polymerization to Generate Analogs of Sulfated Glycoproteins

Sulfated saccharides are critical constituents of many fundamental biological processes. For example, the selectins, proteins that facilitate the recruitment of white blood cells to the blood vessel wall in inflammation, bind several densely O-glycosylated, sulfated proteins [56]. The saccharide epitopes that bind to the selectins are negatively charged, as they possess anionic substituents including sialic acid and/or sulfate groups, and these serve as a basis for monomer design. The stability of the ruthenium carbene initiator to this type of functionality had not been explored. Thus, a first goal was to determine whether ruthenium initiators could facilitate the polymerization of sulfated carbohydrate derivatives, and to this end, monomers containing simple sulfated mimics of natural selectin ligands were designed. The initial monomers that were constructed possessed 3-sulfogalactose or 3,6-disulfogalactose epitopes; these saccharides share the charge distribution of the two putative L-selectin binding units: sulfatides and epitopes from the glycoprotein GlyCAM-1 (Fig. 18A–C). An oxanorbornene or norbornene acid derivative with a short linker attaching the carbohydrate was chosen as the template [45, 46].

The initial polymerizations were conducted with both homogeneous and emulsion conditions. In the homogeneous reactions, a mixture of methanol:dichloroethane was used to dissolve the sulfated monomer, and the initiator was introduced as a solution in dichloroethane. These reactions did not proceed to completion, even upon heating. During the reaction, the formation of a precipitate was observed, which could be attributed to the growing polymer chain. Although the monomer is soluble in methanol as a triethylammonium salt, the

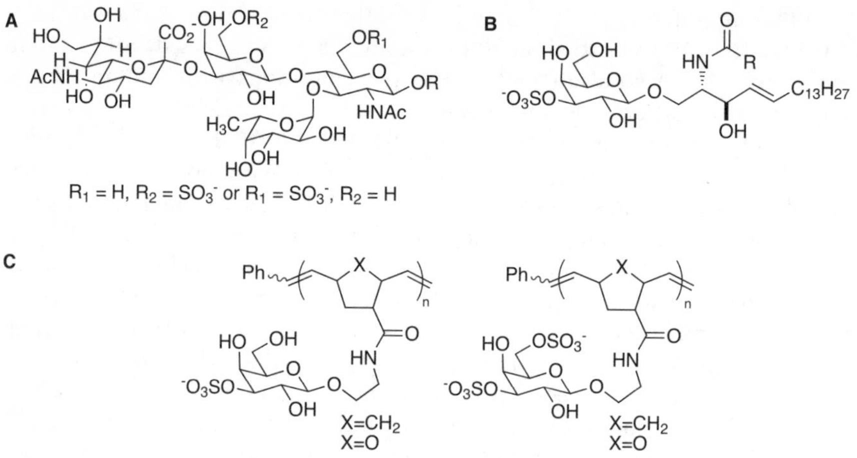

Fig. 18. Sulfated carbohydrates were polymerized under emulsion conditions using the defined ruthenium catalyst to provide glycoprotein mimics

product polymers, which are significantly more polar, are soluble only in aqueous solution. These results demonstrate that multivalent sulfated saccharide derivatives can be synthesized by ROMP, but new conditions were sought that might be amenable to living polymerization. The application of emulsion conditions, using low concentrations of DTAB detergent (1.6 equivalents) to facilitate product isolation, resulted in a rapid almost complete reaction of the monomer.

The multivalent sulfated polymers were evaluated for their ability to inhibit the binding of the selectins to their ligands. The assay involved monitoring the binding of a human leukemia cell line (HL60 cells), which display selectin ligands on their cell surfaces, to immobilized recombinant L, P, or E-selectin (Fig. 19). The monomers did not display any activity, but the polymers were highly effective inhibitors (IC_{50} values as low as 0.084 mM). Interestingly, the polymers produced under homogeneous conditions were more potent inhibitors of all three selectins than those generated under emulsion conditions. The differences in potency may be due to variations in the average lengths of the polymers that are generated under the distinct reaction conditions. In the case of the homogeneous polymerization reactions, the high polarity of the monomers may diminish the rates of initiation and/or propagation, leading to products of broad molecular weight distributions. Longer polymers may exhibit increased potency in bioassays, as was observed with the ConA system. Polymer size influences activity, but the carbohydrate epitope is also critical. The differences between saccharide epitopes are most striking for the emulsion polymers. While the 3-sulfogalactose emulsion polymer shows weak inhibition of P-selectin, the 3,6-disulfogalactose emulsion polymer is a selective, potent, P-selectin inhibitor. The 3,6-disulfogalactose emulsion polymer has an IC_{50} value 500-fold lower than sLe^x, a naturally occurring monovalent ligand. Overall the results of this as-

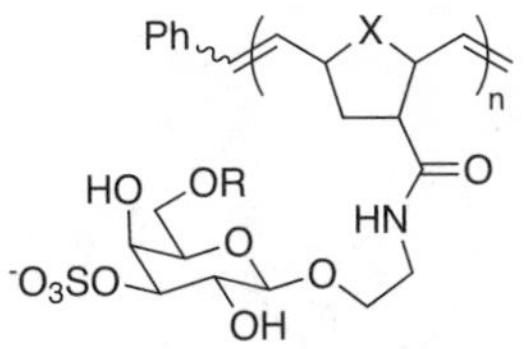

	Compound	P-selectin $(IC_{50}$ mM)	L-selectin $(IC_{50}$ mM)	E-selectin $(IC_{50}$ mM)
A	X=O, R=H	2.2	75% @ 3.0	2.9
	X=O, R=OSO$_3^-$	0.084	1.7	90% @ 3.0
	sLe^x	3.4 ± 0.27	3.5 ± 0.18	3.3 ± 0.17

	Compound	P-selectin $(IC_{50}$ mM)	L-selectin $(IC_{50}$ mM)	E-selectin $(IC_{50}$ mM)
B	X=O, R=H	7.8	0% @ 20.0	20
	X=CH$_2$, R=H	1.2	0% @ 20.0	0% @ 20.0
	X=O, R=OSO$_3^-$	13% @20.0	13% @20.0	13% @20.0
	X=CH$_2$, R=OSO$_3^-$	0.17	18	58% @ 20.0

Fig. 19. The sulfated glycoprotein mimics were tested for biological activity against the selectin family of proteins, which are involved in the inflammatory response. These results
suggest that length may be a factor in the selectivity of polymers for different proteins

say indicate that selective inhibitors can be generated by modifying not only the
carbohydrate epitope but also the length of the polymer.

3.10
Use of Sulfated Polymers to Cause Novel Biological Changes

A recent application of multivalent ligands, which were generated by ROMP, suggests new strategies for manipulating cell surfaces. To date, most of the biological activities of the multidentate ligands created by ROMP have focused on the
synthesis of materials for investigating and inhibiting multivalent protein–saccharide binding events. This study indicates that multivalent ligands can be used
to cause clustering of specific receptors at the cell surface, and this clustering can
lead to the proteolytic release of the target receptor into circulation. The approach may lead to a new method for regulating the presence of proteins at the
cell surface [57].

The interest in using saccharide-substituted polymers to bind and cluster cell
surface proteins arises from studies of L-selectin, a protein involved in inflammation. L-selectin binds glycoproteins that display complex carbohydrates, and

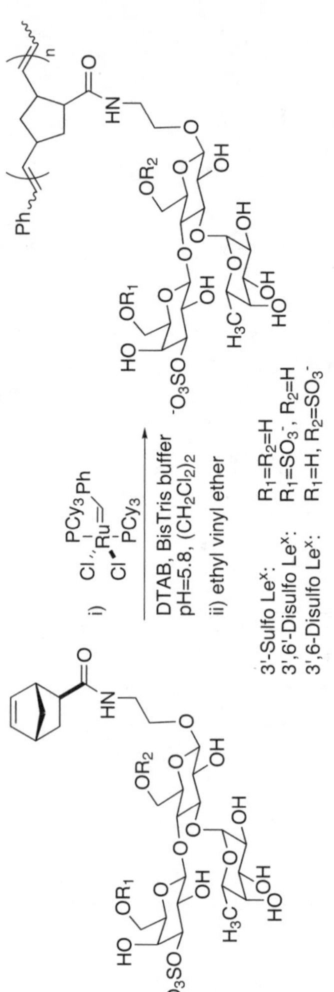

Fig. 20. More complex sulfated polymers were generated and found to have novel effects on the topography of cell surfaces. Specifically, the agents induce the proteolytic release of a target protein, L-selectin, from the cell surface

glycoprotein-inspired materials have been assembled to explore the effects of multivalency on L-selectin recognition. These carbohydrates, 3'-sulfoLe^x, 3',6'-disulfo Le^x, and 3',6-disulfo Le^x, when displayed in a multivalent array more accurately mimic physiological selectin ligands. A synthesis of the trisaccharide-substituted monomers was developed. These monomers could be polymerized, using the emulsion conditions developed for the sulfated galactose polymers, to afford complex polymeric glycoprotein analogs in a single step (Fig. 20). Future studies relying on the production of water soluble materials of this type will be greatly facilitated by the recent development of a water-soluble ruthenium carbene initiator [54, 58].

L-selectin can be proteolytically released from the surface of white blood cells and this release occurs under physiological conditions [59]. A soluble form of L-selectin is found in human blood, but higher concentrations are associated with some disease states. It is not known how the proteolysis of L-selectin is regulated, but circumstantial evidence suggest that binding of L-selectin to glycoprotein mimics might trigger its release. Therefore, the glycoprotein analogs were tested for their ability to cause the cleavage or shedding of L-selectin from the cell surface. The 3',6-disulfo Le^x polymer could effectively cause L-selectin shedding, but the corresponding monomer could not [60]. These studies highlight the ability of the glycoprotein analog displaying 3',6-disulfo Le^x, which has the ability to cluster L-selectin on the cell surface, to induce the proteolytic release of L-selectin from the cell surface. White blood cells that have lost their L-selectin can no longer function in an inflammatory response; therefore, these molecules may lead to the development of new anti-inflammatory strategies.

The glycoprotein mimics, like other selectin inhibitors, compete with natural ligands for selectin binding by non-covalent association. Yet, these agents have an activity that conventional inhibitors do not possess: they can cause a change in covalent bonding (covalent bond cleavage). Ligand-induced shedding irreversibly modifies the cell surface. Because a number of proteins can be released from the cell, these results may serve as a blueprint for the design and synthesis of molecules that can surgically remove unwanted proteins from the cell surface. Thus, access to these glycoprotein analogs through ROMP has facilitated the discovery of a new physiological process, which can be selectively activated. The use of bioactive polymers to modulate the cell surface receptor population is a new horizon for controlling inter- and intracellular recognition and signaling.

4
Conclusion

The synthesis of neobiopolymers using ROMP is a rapidly growing area. Advances in the chemistry of ROMP have led to new strategies for the creation of glycoprotein, peptide, and nucleic acid analogs. In addition, polymers equipped with bioactive groups may be useful for targeted drug delivery or for vaccine development. To date, most researchers exploiting ROMP chemistry have focused on the synthesis of unique polymers with potential biological uses. Significantly, studies of carbohydrate-substituted polymers synthesized by ROMP reveal that these materials can have potent biological activities. They can illuminate underlying mechanisms in biological recognition processes, act as potent inhibitors of physiological recognition events, and facilitate cellular processes including the removal of deleterious proteins on the cell surface. The exciting opportunities for applying tailored polymers to manipulate biological systems depend on new advances in the chemistry of ROMP and an understanding and recognition of potential biological applications. It is this synergy that will lead to the creation of original classes of neobiopolymers and their uses in imaginative applications.

5
References

1. Appella DH, Christianson LA, Klein DA, Powell DR, Huang X, Barchi Jr J J, Gellman SH (1997) Nature 387:381
2. Simon RJ, Kania RS, Zuckermann RN, Huebner VD, Jewell DA, Banville S, Ng S, Wang L, Rosenberg S, Marlowe CA, Spellmeyer DC, Tan R, Frankel AD, Santi DV, Cohen FE, Bartlett PA (1992) Proc Natl Acad Sci USA 89:9367
3. Soth MJ, Nowick JS (1997) Curr Opin Chem Biol 1:120
4. Cho CY, Moran EJ, Cherry SR, Stephans JC, Fodor SPA, Adams CL, Sundaram A, Jacobs JW, Schultz PG (1993) Science 261:1303
5. Klein J, Begli AH (1989) Makromol Chem 190:2527
6. Spaltenstein A, Whitesides GM (1991) J Am Chem Soc 113:686
7. Weigel PH, Schmell E, Lee YC, Roseman S (1978) J Biol Chem 253:330
8. Weigel PH, Schnaar RL, Kuhlenschmidt MS, Schmell E, Lee RT, Lee YC, Roseman S (1979) J Biol Chem 254:10830
9. Billmeyer FWJ (1984) Textbook of Polymer Science, 3rd edn. John Wiley & Sons, New York
10. For a recent review, see Hawker CJ (1997) Acc Chem Res 30:373
11. Ohno K, Tsujii Y, Miyamoto T, Fukuda T, Goto M, Kobayashi K, Akaike T (1998) Macromolecules 31:1064
12. Minoda M, Yamaoka K, Yamada K, Takaragi A, Miyamoto T (1995) Macromol Symp 99:169
13. Yamada K, Minoda M, Miyamoto T (1997) J Polym Sci A: Polym Chem 35:751
14. Yamada K, Yamaoka K, Minoda M, Miyamoto T (1997) J Polym Sci A: Polym Chem 35:255
15. Sawamato M, Aoshima S, Higashimura T (1988) Mackromol Chem, Macromol Symp 13/14:513
16. Sawamato M (1991) Prog Polym Sci 16:111
17. Nomura K, Schrock RR (1996) Macromolecules 29:540
18. Fraser C, Grubbs RH (1995) Macromolecules 28:7248
19. Aoi K, Suzuki H, Okada M (1992) Macromolecules 25:7073
20. Aoi K, Tsutsumiuchi K, Okada M (1994) Macromolecules 27:875
21. Tsutsumiuchi K, Aoi K, Okada M (1997) Macromolecules 30:4013
22. Imanishi Y (1984) . N-Carboxyanhydrides. In: Ivin K, Saegusa T (eds) Ring-Opening Polymerization, vol 2, Chap. 8. Elsevier Applied Science Publishers, New York, p 523
23. Nguyen ST, Johnson LK, Grubbs RH (1992) J Am Chem Soc 114:3974
24. For more information regarding this mechanism, see Ivin KJ, Mol JC (1997) Olefin Metathesis and Metathesis Polymerization. Academic Press, London
25. Grubbs RH (1994) J M S Pure Appl Chem A31:1829
26. Mortell KH, Gingras M, Kiessling LL (1994) J Am Chem Soc 116:12053
27. Schuster MC, Mortell KH, Hegeman AD, Kiessling LL (1997) J Mol Catal A: Chem 116:209
28. Novak BM, Grubbs RH (1988) J Am Chem Soc 110:7542
29. Lee Y, Lee R (1995) Acc Chem Res 28:321
30. Bittiger H (1976) Concanavalin A as a Tool. John Wiley & Sons, New York
31. Weatherman RV, Kiessling LL (1996) J Org Chem 61:534
32. Lynn DM, Kanaoka S, Grubbs RH (1996) J Am Chem Soc 118:784
33. Bazan GC, Khosravi E, Schrock RR, Feast WJ, Gibson VC, O'Regan MB, Thomas JK, Davis WM (1990) J Am Chem Soc 112:8378
34. Feast WJ, Gibson VC, Marshall EL (1992) J Chem Soc, Chem Commun:1157
35. Oskam JH, Schrock RR (1992) J Am Chem Soc 114:7588
36. McConville DH, Wolf JR, Schrock RR (1993) J Am Chem Soc 115:4413
37. O'Dell R, McConville DH, Hofmeister GE, Schrock RR (1994) J Am Chem Soc 116:3414

38. Albagli D, Bazan GC, Schrock RR, Wrighton MS (1993) J Am Chem Soc 115:7328
39. Coles MP, Gibson VC, Mazzariol L, North M, Teasdale WG, Williams CM, Zamuner D (1994) J Chem Soc, Chem Commun:2505
40. Biagini SCG, Bush SM, Gibson VC, Mazzariol L, North M, Teasdale WG, Williams CM, Zagotto G, Zamuner D (1995) Tetrahedron 51:7247
41. Biagini SCG, Coles MP, Gibson VC, Giles MR, Marshall EL, North M (1998) Polymer 39:1007
42. For a recent review, see: DesMesmaeker A, Härer R, Martin P, Moser HE (1995) Acc Chem Res 28:366
43. Gibson VC, Marshall EL, North M, Robson DA, Williams PJ (1997) J Chem Soc, Chem Commun:1095
44. Biagini SCG, Gibson VC, Giles MR, Marshall EL, North M (1997) J Chem Soc, Chem Commun:1097
45. Manning DD, Strong LE, Hu X, Beck PJ, Kiessling LL (1997) Tetrahedron 53:11937
46. Manning DD, Hu X, Beck P, Kiessling LL (1997) J Am Chem Soc 119:3161
47. Mortell KH, Weatherman RV, Kiessling LL (1996) J Am Chem Soc 118:2297
48. Kanai M, Mortell KH, Kiessling LL (1997) J Am Chem Soc 119:9931
49. Page MI, Jencks WP (1971) Proc Natl Acad Sci USA 68:1678
50. Jencks WP (1981) Proc Natl Acad Sci USA 78:4046
51. Crothers DM, Metzger H (1972) Immunochemistry 9:341
52. Kanai M, Kiessling LL, unpublished results
53. Dias EL, Nguyen ST, Grubbs RH (1997) J Am Chem Soc 119:3887
54. Lynn DM, Mohr B, Grubbs RH (1998) J Am Chem Soc 120:1627
55. Kurutz JW, Kiessling LL, unpublished results
56. Kansas GS (1996) Blood 88:3259
57. Kiessling LL, Gordon EJ (1998) J Chem Biol 5:R49
58. Mohr B, Lynn DM, Grubbs RH (1996) Organometallics 15:4317
59. Kishimoto TK, Jutila MA, Berg EL, Butcher, EC (1989) Science 245:1238
60. Gordon EJ, Sanders WJ, Kiessling LL (1998) Nature 392:30

Printing: Mercedesdruck, Berlin
Binding: Buchbinderei Lüderitz & Bauer, Berlin